Quantum Communication—Celebrating the Silver Jubilee of Teleportation

Quantum Communication—Celebrating the Silver Jubilee of Teleportation

Editors

Rotem Liss
Tal Mor

MDPI • Basel • Beijing • Wuhan • Barcelona • Belgrade • Manchester • Tokyo • Cluj • Tianjin

Editors
Rotem Liss
Technion–Israel Institute of Technology
Israel

Tal Mor
Technion–Israel Institute of Technology
Israel

Editorial Office
MDPI
St. Alban-Anlage 66
4052 Basel, Switzerland

This is a reprint of articles from the Special Issue published online in the open access journal *Entropy* (ISSN 1099-4300) (available at: https://www.mdpi.com/journal/entropy/special_issues/Quantum_Communication).

For citation purposes, cite each article independently as indicated on the article page online and as indicated below:

LastName, A.A.; LastName, B.B.; LastName, C.C. Article Title. *Journal Name* **Year**, *Article Number*, Page Range.

ISBN 978-3-03943-026-0 (Hbk)
ISBN 978-3-03943-027-7 (PDF)

© 2020 by the authors. Articles in this book are Open Access and distributed under the Creative Commons Attribution (CC BY) license, which allows users to download, copy and build upon published articles, as long as the author and publisher are properly credited, which ensures maximum dissemination and a wider impact of our publications.

The book as a whole is distributed by MDPI under the terms and conditions of the Creative Commons license CC BY-NC-ND.

Contents

About the Editors vii

Rotem Liss and Tal Mor
Quantum Communication—Celebrating the Silver Jubilee of Teleportation
Reprinted from: *Entropy* **2020**, *22*, 628, doi:10.3390/e22060628 1

Francesco De Martini and Fabio Sciarrino
Twenty Years of Quantum State Teleportation at the Sapienza University in Rome
Reprinted from: *Entropy* **2019**, *21*, 768, doi:10.3390/e21080768 9

Nicolas Gisin
Entanglement 25 Years after Quantum Teleportation: Testing Joint Measurements in Quantum Networks
Reprinted from: *Entropy* **2019**, *21*, 325, doi:10.3390/e21030325 25

Gilles Brassard, Luc Devroye and Claude Gravel
Remote Sampling with Applications to General Entanglement Simulation
Reprinted from: *Entropy* **2019**, *21*, 92, doi:10.3390/e21010092 37

William K. Wootters
A Classical Interpretation of the Scrooge Distribution
Reprinted from: *Entropy* **2018**, *20*, 619, doi:10.3390/e20080619 55

Michel Boyer, Rotem Liss and Tal Mor
Attacks against a Simplified Experimentally Feasible Semiquantum Key Distribution Protocol
Reprinted from: *Entropy* **2018**, *20*, 536, doi:10.3390/e20070536 73

Kan Wang, Xu-Tao Yu, Xiao-Fei Cai and Zai-Chen Zhang
Probabilistic Teleportation of Arbitrary Two-Qubit Quantum State via Non-Symmetric Quantum Channel
Reprinted from: *Entropy* **2018**, *20*, 238, doi:10.3390/e20040238 83

About the Editors

Rotem Liss received his B.Sc. (2015) and M.Sc. (2017) in Computer Science from the Technion–Israel Institute of Technology, where he is currently (2020) finalizing his Ph.D. He is a recipient of the Daniel Fellowship (2017) and the Jacobs Fellowship (2019).

Rotem's main research interests are focused on quantum cryptography and quantum entanglement. His scientific contributions include the security analysis of several quantum key distribution protocols and the analysis of entanglement and topological order inside a Bloch sphere.

Tal Mor, Ph.D., has been a professor at the Computer Science Department of the Technion–Israel Institute of Technology since 2002. He received his M.Sc. from Tel Aviv University (in Yakir Aharonov's group), and his Ph.D. from the Technion, supervised by Asher Peres and Eli Biham. He was previously a postdoctoral fellow at the University of Montreal and the University of California, Los Angeles (UCLA).

Tal's most prominent scientific contributions include the "photon-number splitting" attack against quantum key distribution protocols, semi-quantum key distribution, the first variant of "measurement-device-independent" quantum key distribution (MDI-QKD), the "algorithmic cooling" method for spins, and quantum nonlocality without entanglement.

Editorial

Quantum Communication—Celebrating the Silver Jubilee of Teleportation

Rotem Liss * and Tal Mor

Computer Science Department, Technion, Haifa 3200003, Israel; talmo@cs.technion.ac.il
* Correspondence: rotemliss@cs.technion.ac.il

Received: 29 May 2020; Accepted: 2 June 2020; Published: 6 June 2020

Keywords: quantum communication; quantum teleportation; quantum entanglement

---

1. Introduction: Quantum Teleportation—Meaning and Influence

In 1993, Charles H. Bennett, Gilles Brassard, Claude Crépeau, Richard Jozsa, Asher Peres, and William K. Wootters published their seminal paper presenting quantum teleportation, titled "Teleporting an unknown quantum state via dual classical and Einstein–Podolsky–Rosen channels" [1]. Their paper presents and answers the question "Can we transmit an unknown quantum state *without physically sending it*?" Namely, can we send enough information about our unknown quantum state, in a way that would enable the receiver to obtain (i.e., regenerate) it? Their paper provides a striking answer: "Yes. An arbitrary state of a quantum bit (denoted by $|\psi\rangle \triangleq \cos\left(\frac{\theta}{2}\right)|0\rangle + e^{i\phi}\sin\left(\frac{\theta}{2}\right)|1\rangle$) can be transmitted *if* both the sender and the receiver share a maximally entangled quantum state (for example, the *singlet* state, denoted by $|\Psi^-\rangle \triangleq \frac{|01\rangle - |10\rangle}{\sqrt{2}}$) *and* the sender can transmit classical messages (only two standard/classical bits) to the receiver". This answer, which presented the *quantum teleportation* protocol, has revolutionized the field of quantum communication.

Intuitively, the teleportation paper proves the equivalence "a quantum communication channel = a shared entangled state + a classical communication channel". In particular, "*sending* an unknown state of **one** *quantum* bit can be done by *sharing* (ahead of time) **one** maximally entangled state of two *quantum* bits + *sending* **two** *classical* bits". The above equivalence is very important, because quantum channels tend to be much less reliable (and much more prone to losses and errors) than classical channels; moreover, even if the sender and the receiver only share (many) *noisy* entangled states, they can still employ quantum teleportation by first distilling (a fewer number of) nearly *maximally* entangled states [2]. (This method, in particular, makes it possible to transmit arbitrarily faithful quantum states over a *noisy* quantum channel [2], even without using quantum error-correcting codes [3]).

To see how surprising this result is, let us represent the quantum bit as an *arrow* directed at some arbitrary direction in the three-dimensional space (see Figure 1 for a two-dimensional illustration). The arrow's direction can be represented in spherical coordinates by the two angles θ, ϕ (note that θ, ϕ are also the two angles that appear in the mathematical representation $|\psi\rangle \triangleq \cos\left(\frac{\theta}{2}\right)|0\rangle + e^{i\phi}\sin\left(\frac{\theta}{2}\right)|1\rangle$ of the quantum bit). Therefore, the corresponding *classical* question is "Can we transmit the arrow's direction *without physically sending the arrow*?" The obvious classical answer is "Yes, but only if we send the real numbers θ, ϕ". Namely, in the classical case, even when the sender *knows* the arrow's direction, a very large number of classical bits must be sent so that the receiver can reconstruct the approximate direction of the arrow (the degree of precision dictates the number of sent bits; infinite precision requires an *infinite* number of bits). On the other hand, *two* classical bits would give us a very limited amount of information, not allowing the receiver to recover the arrow's direction at any reasonable amount of precision. This is true even if the sender and the receiver share some information in advance, assuming that the arrow's direction is chosen randomly and independently of the shared information. (In a limited classical case, where the sender wants the receiver to get the probability distribution of *one* biased coin, we can have some kind of "classical teleportation", even if that distribution is unknown to the sender; see details in [4]).

Figure 1. We illustrate the power of quantum teleportation by representing the quantum bit as an *arrow* (a two-dimensional arrow in this drawing; a three-dimensional arrow in general) inside a unit sphere. In the general three-dimensional case, this representation is known as the *Bloch sphere representation*. The sender would like to transmit the arrow's direction to the receiver, without physically sending the arrow.

The *quantum* case seems even worse: if the sender holds the unknown quantum state $|\psi\rangle$ and wants to transmit it to the receiver, the sender apparently still has to send the two real numbers θ, ϕ. Moreover, due to the peculiar properties of quantum mechanics, those real numbers are now *not even known* to the sender, because the description of $|\psi\rangle \triangleq \cos\left(\frac{\theta}{2}\right)|0\rangle + e^{i\phi}\sin\left(\frac{\theta}{2}\right)|1\rangle$ is unknown to the sender. (Note that the sender cannot discover the description of $|\psi\rangle$, and any attempt to do so would irreversibly damage the quantum state). Nonetheless, the quantum teleportation paper proves that by using the extraordinary power of quantum entanglement, *only two* classical bits need to be sent.

The teleportation paper is one of the most prominent examples of the counterintuitive power of quantum communication; other notable examples include quantum cryptography [5,6], violations of Bell's inequality [7], and even the basic phenomena of quantum entanglement and EPR pairs [8].

2. The Discovery of Quantum Teleportation: History, Notes, and Stories

Like any groundbreaking result, there are several interesting stories surrounding the discovery of quantum teleportation. Perhaps most interesting of all is the story of the actual invention of quantum teleportation, as recounted by Gilles Brassard and printed here for the first time (except for an earlier personal account in French [9]):

"It all started in August 1992, when I was attending the annual CRYPTO conference. Charlie Bennett gave me a paper that had appeared in *Physical Review Letters* one year earlier, saying 'I think this will interest you'. Right he was! That was the paper by Asher Peres and William (Bill) K. Wootters [10] in which they considered the following problem: if two participants hold identical copies of an unknown quantum state $|\psi\rangle$, so that the state of their joint system is $|\psi\rangle_A \otimes |\psi\rangle_B$, how much information can they discover about $|\psi\rangle$ if they are restricted to local quantum operations and classical communication (this is of course what became known later as LOCC)? In that paper, Peres and Wootters studied so-called ping-pong protocols in which more information can be obtained by increasing the number of interaction rounds, but they were unable to get quite as much information as if the two identical quantum states were in the same location, enabling the possibility of a joint measurement. Their paper left open the following question: can LOCC measurements provide as much information as joint measurements?

At the time, I had never met Peres or Wootters, and in fact I had never heard of them. I met them both a few months later by a pleasant coincidence, at the October 1992 *Workshop on Physics and Computation* held in Dallas. After discussing the paper with its authors, I invited

Bill to come to Montréal to give a talk about it the following month. Somehow, I had a feeling this would be momentous, and therefore I invited Claude Crépeau (who was in Paris at the time) and Charlie Bennett to attend the talk at my expense. Richard Jozsa was in the audience as well because he was my research assistant at the time. After Bill explained the conundrum, Charlie raised his hand and asked an apparently inane question: 'What difference would it make if the two participants shared an EPR pair?' (that's what we called entanglement in those days). Not surprisingly, Bill replied 'I don't know!' and then went on with his talk. Immediately afterwards, we all moved to my office and brainstormed about Charlie's question. By the next morning, the answer was clear: in the presence of entanglement, one party teleports $|\psi\rangle$ to the other, who then performs the optimal joint measurement. It is fair to say that we were able to invent quantum teleportation within less than 24 hours because none of us was trying to achieve this obviously impossible task! Of course, we realized that this invention was far more important than the solution it offered to the problem at hand, but I don't think any of us anticipated how important it would become. We quickly invited Asher to join the collaboration and, within eleven days, the paper was submitted to *Physical Review Letters*. The rest is history".

The writing process of that seminal paper was not exempt from dilemmas. Gilles Brassard describes one of them—the length of the paper:

"Once we had a version of the paper that we really liked, we noticed that it was just a little too long for the then strict limit of four pages imposed by *Physical Review Letters*. We could not find anything that we would be comfortable leaving out. That was when a devilish idea came to me. Given that the type is smaller in figure captions (8.5 points) than in the main text (9.5 points), why not squeeze in some content there? We relegated the proof that successful teleportation of one qubit *requires* the transmission of two classical bits to what became a 27-line caption for Figure 2 (see [1]), which saved exactly the required amount of space to fit the paper snuggly in four pages. Ironically, we ended up being the first paper of its issue, and the space needed for the journal header made us spill on a fifth page!"

Another important dilemma was the order of the authors' names. Readers unfamiliar with the advantages and disadvantages of alphabetical order may not be able to understand and appreciate the subtleties of the following story. Alphabetical order for authors' names is customary in our field, in contrast to the "contribution order", which is conventional in many others. There are researchers who participate only or mostly in alphabetical-order papers, and it is very important to many of them to *avoid combining the two methods*: combining in that way could have a negative potential impact on both their own research career and the careers of their alphabetic co-authors.

The original teleportation paper listed authors in alphabetical order. Charles Bennett, who frequently held the position of first author due to his last name, with many papers being cited as "Bennett et al". felt he was being over-credited in the eyes of those accustomed to contribution order. For this reason, at some point before submission of the teleportation paper, he suggested the use of reverse alphabetical order for the authors, which would have placed Bill Wootters as first author. This idea was almost immediately rejected by Wootters himself. Gilles Brassard, who has never once strayed from alphabetical order throughout his entire career, told us years later that he felt so strongly about this issue, that he would have withdrawn his name from the paper had Bennett's reverse authorship idea been carried out. Of course he could not have known this at the time, but taking himself off the paper might have prevented him from sharing the Wolf Prize with Charles Bennett one quarter of a century later. (Details about the Wolf Prize won by Bennett and Brassard, which they received *both* for quantum cryptography and quantum teleportation, are provided at the end of the current section).

Yet another dilemma was the *name* of the new method. Asher Peres objected to the original name "teleportation" because it mixed the Greek prefix "tele-" with the Latin-based root "port".

Peres suggested the alternative name "telepheresis", but the other authors disagreed, so the name remained "teleportation".

The quantum teleportation paper received excellent reviews before being accepted to *Physical Review Letters*. One of the reviewers, N. David Mermin, described the paper as a "charming, readable, thought-provoking paper", and predicted that "this novel method [...] will become an important one of the intellectual tools available to anybody [...]" (see Mermin's paper [11], where he disclosed his full referee report on the quantum teleportation paper). Finally, the paper was published on 29 March 1993 [1] in *Physical Review Letters*, profoundly advancing the field of quantum communication and bringing new researchers to the fast-evolving field of quantum information processing. In particular, it influenced Tal Mor, who is one of the authors of this editorial, as he describes below.

> When the teleportation paper was published (1993), I was an M.Sc. student in Tel Aviv University (Israel) in the group led by Yakir Aharonov, together with Sandu Popescu (who was a Ph.D. student at the time) and Lev Vaidman (who was a postdoctoral researcher). All three of us (Popescu, Vaidman, and I) were extremely excited about the teleportation paper: Popescu suggested a method [12] leading to the first experimental realization of quantum teleportation [13] (note that quantum teleportation was experimentally demonstrated in 1997–1998 by three research groups [13–15]); Vaidman suggested teleportation of continuous quantum variables [16], leading to a theoretical extension [17] and its experimental realization [15]; and I decided to start my Ph.D. with Asher Peres, who was one of the teleportation paper's authors. Although I concentrated on quantum cryptography, I also gave a lot of thought to quantum teleportation: I presented teleportation as a special case of POVM (generalized measurements) in my first talk at an international conference [18–20], and I suggested how a classical variant of teleportation could look like (a concept I published years later [4]).

> The quantum teleportation paper and its experimental realizations intrigued not only scientists, but also media reporters. When one of them asked Peres whether quantum teleportation teleports only a person's body, or also the soul, Peres answered that it teleports *only* the soul [21]—a funny, thought-provoking reply from someone like Peres, who enjoyed describing himself as a devout atheist!

> During my Ph.D. and postdoctoral research, I became acquainted with all six authors of the teleportation paper. I even asked them to autograph an original reprint of the paper—so I now own the only copy of the quantum teleportation paper signed by all six co-authors! (Admittedly, it was pretty hard to obtain this signed copy. Unfortunately, Gilles Brassard, who was the last co-author to sign, lost the copy signed by the five other co-authors in his office; later, he sent me an e-mail including the "good news"—that he found the signed copy of the teleportation paper—and the "bad news"—that he lost it again; finally, he found it *again*, signed it, and immediately mailed it from Canada to me in Israel, and I received it. Then, *I* lost it in *my* office... where I may find it again some day).

> Subsequently, I had two opportunities to celebrate the quantum teleportation paper and honor some of its authors at my institution (Technion, Haifa, Israel): when I organized the QUBIT 2003 conference, celebrating 10 years of quantum teleportation, with Asher Peres as the guest of honor [22]; and when I organized the QUBIT 2018 conference, celebrating the Wolf Prize of Charles Bennett and Gilles Brassard, with both of them as the keynote speakers [23].

Charles Bennett and Gilles Brassard won the 2018 Wolf Prize in physics "for founding and advancing the fields of Quantum Cryptography and Quantum Teleportation". The jury panel acknowledged the enormous importance of quantum teleportation:

> "In the 1990's they [Bennett and Brassard], together with four colleagues, invented quantum teleportation which allows the communication of quantum information over classical

channels, also a task previously believed to be impossible. Two decades after their proposal, quantum teleportation has now been demonstrated over distances exceeding 1000 kilometers and is clearly destined to play a major role in future secure communications". [24]

3. The Papers in This Special Issue

This special issue is dedicated to celebrating the silver jubilee of the seminal teleportation paper, and it features contributions from various areas of quantum communication.

Francesco De Martini and Fabio Sciarrino, in their paper "Twenty years of quantum state teleportation at the Sapienza University in Rome" [25], review various experiments of quantum teleportation that were conducted at the Sapienza University in Rome, ranging from the *first* teleportation experiment (1997) to several variations and generalizations of teleportation, such as active teleportation and quantum machines based on teleportation.

Nicolas Gisin, in his paper "Entanglement 25 years after quantum teleportation: Testing joint measurements in quantum networks" [26], discusses quantum entanglement from an unusual perspective: that of entangled *measurements* rather than entangled *states*. In particular, Gisin raises the question of whether entangled measurements can be used for generating non-classical output correlations in various quantum networks, and suggests a few candidates that may present such non-classical correlations.

Gilles Brassard, Luc Devroye, and Claude Gravel, in their paper "Remote sampling with applications to general entanglement simulation" [27], provide a (classical) sampling scheme: their scheme allows the user to sample *exactly* from a discrete probability distribution when the defining parameters of this probability distribution are partitioned between several remote parties. Furthermore, they apply their sampling scheme to the classical simulation of quantum entanglement measurements in the most general scenario, and analyze its communication complexity.

William K. Wootters, in his paper "A classical interpretation of the Scrooge distribution" [28], shows how to derive a special *quantum* ensemble of pure states, known as the "Scrooge ensemble" (or "Scrooge distribution"), from a *classical* communication scenario. Specifically, he proves that a real-amplitude variant of the Scrooge distribution naturally appears in a classical communication scheme, and that the standard (complex-amplitude) Scrooge distribution appears in a modified version of the same communication scheme.

Michel Boyer, Rotem Liss, and Tal Mor, in their paper "Attacks against a simplified experimentally feasible semiquantum key distribution protocol" [29], explore the security of a semiquantum key distribution (SQKD) protocol that seems easy to implement in practice. In particular, they analyze a simplified variant of the previously published "Mirror" SQKD protocol, and prove that unlike the original Mirror protocol (which was proved completely robust), its simplified variant is completely insecure if the tolerated loss rate is high.

Kan Wang, Xu-Tao Yu, Xiao-Fei Cai, and Zai-Chen Zhang, in their paper "Probabilistic teleportation of arbitrary two-qubit quantum state via non-symmetric quantum channel" [30], propose a variant of quantum teleportation: their scheme allows teleporting an arbitrary two-qubit state from Alice to Bob, given that Alice and Bob share one partially entangled pure three-qubit state and one partially entangled pure two-qubit state. Their teleportation scheme is probabilistic and unambiguous: namely, it may fail with constant probability, but the users know whether it succeeded or failed.

We hope that the papers in this special issue give insight regarding the different areas of quantum communication—most notably, quantum teleportation, quantum entanglement, and quantum cryptography.

Funding: The work of T.M. and R.L. was supported in part by the Israeli MOD.

Acknowledgments: We thank all the authors who submitted their contributions to this special issue. We acknowledge all the anonymous reviewers and the editorial staff of Entropy for their ongoing support.

We also thank Gilles Brassard for his invaluable help regarding all parts of this editorial and for writing his historical account on the invention of quantum teleportation.

Conflicts of Interest: The authors declare no conflict of interest.

References

1. Bennett, C.H.; Brassard, G.; Crépeau, C.; Jozsa, R.; Peres, A.; Wootters, W.K. Teleporting an unknown quantum state via dual classical and Einstein–Podolsky–Rosen channels. *Phys. Rev. Lett.* **1993**, *70*, 1895–1899. [CrossRef] [PubMed]
2. Bennett, C.H.; Brassard, G.; Popescu, S.; Schumacher, B.; Smolin, J.A.; Wootters, W.K. Purification of noisy entanglement and faithful teleportation via noisy channels. *Phys. Rev. Lett.* **1996**, *76*, 722–725. [CrossRef] [PubMed]
3. Shor, P.W. Scheme for reducing decoherence in quantum computer memory. *Phys. Rev. A* **1995**, *52*, R2493–R2496. [CrossRef] [PubMed]
4. Mor, T. On classical teleportation and classical non-locality. *Int. J. Quantum Inf.* **2006**, *4*, 161–171. [CrossRef]
5. Bennett, C.H.; Brassard, G. Quantum cryptography: Public key distribution and coin tossing. In Proceedings of the International Conference on Computers, Systems & Signal Processing, Bangalore, India, 10–12 December 1984; pp. 175–179.
6. Bennett, C.H.; Brassard, G. Quantum cryptography: Public key distribution and coin tossing. *Theor. Comput. Sci.* **2014**, *560*, 7–11, doi:10.1016/j.tcs.2014.05.025. [CrossRef]
7. Bell, J.S. On the Einstein Podolsky Rosen paradox. *Phys. Phys. Fiz.* **1964**, *1*, 195–200. [CrossRef]
8. Einstein, A.; Podolsky, B.; Rosen, N. Can quantum-mechanical description of physical reality be considered complete? *Phys. Rev.* **1935**, *47*, 777–780. [CrossRef]
9. Brassard, G.; Crépeau, C. « Nous avons inventé la téléportation en un jour ! » [English: "We invented teleportation in one day!"]. *La Rech.* **2005**, *386*, 32–34.
10. Peres, A.; Wootters, W.K. Optimal detection of quantum information. *Phys. Rev. Lett.* **1991**, *66*, 1119–1122. [CrossRef]
11. Mermin, N.D. Copenhagen computation: How I learned to stop worrying and love Bohr. *IBM J. Res. Dev.* **2004**, *48*, 53–61. [CrossRef]
12. Popescu, S. An optical method for teleportation. *arXiv* **1995**, arXiv:quant-ph/9501020.
13. Boschi, D.; Branca, S.; De Martini, F.; Hardy, L.; Popescu, S. Experimental realization of teleporting an unknown pure quantum state via dual classical and Einstein-Podolsky-Rosen channels. *Phys. Rev. Lett.* **1998**, *80*, 1121–1125. [CrossRef]
14. Bouwmeester, D.; Pan, J.W.; Mattle, K.; Eibl, M.; Weinfurter, H.; Zeilinger, A. Experimental quantum teleportation. *Nature* **1997**, *390*, 575–579. [CrossRef]
15. Furusawa, A.; Sørensen, J.L.; Braunstein, S.L.; Fuchs, C.A.; Kimble, H.J.; Polzik, E.S. Unconditional quantum teleportation. *Science* **1998**, *282*, 706–709. [CrossRef]
16. Vaidman, L. Teleportation of quantum states. *Phys. Rev. A* **1994**, *49*, 1473–1476. [CrossRef]
17. Braunstein, S.L.; Kimble, H.J. Teleportation of continuous quantum variables. *Phys. Rev. Lett.* **1998**, *80*, 869–872. [CrossRef]
18. Mor, T. TelePOVM—New faces of teleportation. *arXiv* **1996**, arXiv:quant-ph/9608005.
19. Mor, T. Tele-POVM: New faces of teleportation. In *A Golden Jubilee Event of the Tata Institute of Fundamental Research (TIFR) on the Foundation of Quantum Theory*; TIFR: Bombay, India, September 1996.
20. Brassard, G.; Horodecki, P.; Mor, T. TelePOVM—A generalized quantum teleportation scheme. *IBM J. Res. Dev.* **2004**, *48*, 87–97. [CrossRef]
21. Avron, J.E.; Bennett, C.H.; Mann, A.; Wootters, W.K. [In the "Obituaries" section:] Asher Peres. *Phys. Today* **2005**, *58*, 65–66. [CrossRef]
22. QUBIT 2003: Celebrating 10 years of Teleportation. 9–10 April 2003. Available online: http://www.cs.technion.ac.il/~talmo/Qubitconf/QUBIT-2003/ (accessed on 3 June 2020).
23. QUBIT 2018—Quantum Communication: Celebrating Bennett and Brassard's Wolf Prize for Physics. 3 June 2018. Available online: https://cyber.technion.ac.il/qubit-2018-celebrating-bennett-and-brassards-wolf-prize-for-physics/ (accessed on 3 June 2020).

24. The Spokesperson's Office of the President of Israel. *Laureates of 2018 Wolf Prize in Music and Sciences Announced*. February 2018. Available online: https://mfa.gov.il/MFA/PressRoom/2018/Pages/Laureates-of-2018-Wolf-Prize-announced-12-February-2018.aspx (accessed on 3 June 2020).
25. De Martini, F.; Sciarrino, F. Twenty years of quantum state teleportation at the Sapienza University in Rome. *Entropy* **2019**, *21*, 768. [CrossRef]
26. Gisin, N. Entanglement 25 years after quantum teleportation: Testing joint measurements in quantum networks. *Entropy* **2019**, *21*, 325. [CrossRef]
27. Brassard, G.; Devroye, L.; Gravel, C. Remote sampling with applications to general entanglement simulation. *Entropy* **2019**, *21*, 92. [CrossRef]
28. Wootters, W.K. A classical interpretation of the Scrooge distribution. *Entropy* **2018**, *20*, 619. [CrossRef]
29. Boyer, M.; Liss, R.; Mor, T. Attacks against a simplified experimentally feasible semiquantum key distribution protocol. *Entropy* **2018**, *20*, 536. [CrossRef]
30. Wang, K.; Yu, X.T.; Cai, X.F.; Zhang, Z.C. Probabilistic teleportation of arbitrary two-qubit quantum state via non-symmetric quantum channel. *Entropy* **2018**, *20*, 238. [CrossRef]

© 2020 by the authors. Licensee MDPI, Basel, Switzerland. This article is an open access article distributed under the terms and conditions of the Creative Commons Attribution (CC BY) license (http://creativecommons.org/licenses/by/4.0/).

Review

Twenty Years of Quantum State Teleportation at the Sapienza University in Rome

Francesco De Martini and Fabio Sciarrino *

Dipartimento di Fisica—Sapienza Università di Roma, P.le Aldo Moro 5, I-00185 Roma, Italy
* Correspondence: fabio.sciarrino@uniroma1.it

Received: 31 July 2018; Accepted: 14 July 2019; Published: 6 August 2019

Abstract: Quantum teleportation is one of the most striking consequence of quantum mechanics and is defined as the transmission and reconstruction of an unknown quantum state over arbitrary distances. This concept was introduced for the first time in 1993 by Charles Bennett and coworkers, it has then been experimentally demonstrated by several groups under different conditions of distance, amount of particles and even with feed forward. After 20 years from its first realization, this contribution reviews the experimental implementations realized at the Quantum Optics Group of the University of Rome La Sapienza.

Keywords: quantum teleportation; entanglement; photonics; quantum information

1. Introduction

Entanglement is today the core of many key discoveries ranging from quantum teleportation [1], to quantum dense coding [2], quantum computation [3–5] and quantum cryptography [6,7]. Quantum communication protocols such as device-independent quantum key distribution [8] are heavily based on entanglement to reach nonlocality-based communication security [9]. The prototype for quantum information transfer using entanglement as a communication channel is the quantum state teleportation (QST) protocol, introduced for the first time in 1993 by Charles Bennett et al. [1], where a sender and a receiver share a maximally entangled state which they can use to perfectly transfer an unknown quantum state. Undoubtedly, quantum teleportation is one of the most counterintuitive consequences of quantum mechanics and it is defined as the transmission and reconstruction over arbitrary distances of an unknown quantum state.

This protocol represents a milestone in theoretical quantum information science [10–12] and lies at the basis of many technological applications such as quantum communication via quantum repeaters [13,14] or gate teleportation [15]. Experimentally, this protocol has been demonstrated by several groups [16–22]. It has been implemented over hundreds of kilometers in free-space [23,24] and more recently in a ground-to-satellite experiment [25]. Employed platforms include mainly photonic qubits [16–18,20–22,26,27], but also nuclear magnetic resonance [28], atomic ensembles [29,30], trapped atoms [31,32] and solid-state systems [33–35]. The progress from the fundamental and technological point of view has continued, allowing the achievement of teleportation of multiple degrees of freedom of a single photon [27]. A quantum space race has started with the satellite-based distribution of entangled photon pairs to two locations, separated by 1203 kilometers on Earth [36], and the first satellite based quantum teleportation. Quantum teleportation experiments using deployed optical fibres for the distribution of entangled pairs have been also recently reported [37–39]: A key ingredient to build a quantum repeater. Quantum teleportation is a key primitive for quantum information processing, that has also been adopted for fundamental tests of quantum mechanics, such as one of the first loophole free Bell tests where entanglement swapping was exploited [40].

The present manuscript reviews the different experiments related to the teleportation of quantum states carried out in the Department of Physics, Sapienza University of Rome. In Section 2 we will briefly summarize the quantum teleportation protocol and its extension to entangled photon pairs: The so-called entanglement swapping. As the first experimental focus, in Section 3, we will describe the first quantum teleportation experiment performed in Rome. This scheme adopted two photons in order to encode the three qubits evolved in the QST protocol. This implementation had the unique capability to discriminate the four Bell states involved in an Alice node. We will then move to Section 4, where a different experimental approach has been adopted. There all the qubits are encoded in the vacuum-one photon Fock basis. This encoding further simplifies the required apparatus. The following step, described in Section 5, has then be to address a missing key ingredient in all the previous implementations of QST: The unitary operator U_i to be implemented from Bob depending on the Alice outcome. This achievement was accomplished again, exploiting the concept of the vacuum-one photon qubit. This realization completed the contributions on the implementation of the original QST scheme. However, new scenarios arose from the modification of the teleportation. In 2004, we could identify how to modify the QST procedure in order to implement two fundamental optimal quantum machines: the universal NOT gate, and the universal optimal quantum cloning: these Universal Quantum Machines are discussed in Section 6. Finally, the last two sections are Sections 7 and 8 are, respectively, devoted to a brief discussion on how to address the classical-quantum transition exploiting optimal quantum cloners, and on the future perspective of teleportation within quantum networks.

2. Teleportation and Entanglement Swapping

2.1. Quantum Teleportation Protocol

The scheme of the teleportation protocol is depicted in Figure 1. By exploiting one maximally entangled state of the Bell basis, we can perform an experimental protocol of quantum teleportation, where two parties, named respectively Alice and Bob, are involved. The photon A of the entangled pair $|\Psi^-\rangle_{A,B} = \frac{1}{\sqrt{2}}(|01\rangle_{AB} - |10\rangle_{AB})$ is sent to Alice and the other photon B to Bob in order to share an entangled state. Let us consider that an unknown quantum state $|\Psi\rangle_T = \alpha|0\rangle_T + \beta|1\rangle_T$ was sent to Alice in order to be teleported, then the state of the whole system can be written as:

$$|\Psi\rangle_T \otimes |\Psi^-\rangle_{A,B} = -\frac{1}{2}|\Psi^-\rangle_{T,A}(\alpha|0\rangle_B + \beta|1\rangle_B) \quad (1)$$

$$-\frac{1}{2}|\Psi^+\rangle_{T,A}(\alpha|0\rangle_B - \beta|1\rangle_B) \quad (2)$$

$$+\frac{1}{2}|\Phi^-\rangle_{T,A}(\alpha|1\rangle_B + \beta|0\rangle_B) \quad (3)$$

$$+\frac{1}{2}|\Phi^+\rangle_{T,A}(\alpha|1\rangle_B - \beta|0\rangle_B). \quad (4)$$

At this point in order to transfer the unknown state to Bob's particle we must perform a Bell state measurement (BSM) in Alice's station. The BSM is a projective measurement in the Bell basis, able to discriminate among the four two-mode entangled states. After the measurement performed by Alice, the state on Bob's side is $U_i|\Psi\rangle_T$. The result of the measurement performed by Alice must be communicated to Bob by means of a classical channel and, according to this one, Bob applies the appropriate Pauli operations (σ_x, σ_z, σ_y or nothing) on his station to complete the teleportation process and retrieve the input state since $U_i^\dagger U_i |\Psi\rangle_T = |\Psi\rangle_T$. Without the communication of the BSM result the particle B would be described by a fully mixed state.

2.2. Entanglement-Swapping Protocol

Is it possible to get entanglement between particles which have never interacted in the past? This simple question has motivated many theoretical and experimental works [1,41,42]. In order to better explain this protocol we will need four parties: Alice, Bob, Victor and Thomas. Let Alice share a maximally entangled state $|\Phi^+\rangle_{AB} = \frac{1}{\sqrt{2}}(|00\rangle_{AB} + |11\rangle_{AB})$ with Bob while Victor and Thomas share the same state between them $|\Phi^+\rangle_{VT}$. At this point the state of the whole system can be written as:

$$|\Phi^+\rangle_{AB} \otimes |\Phi^+\rangle_{VT}. \tag{5}$$

The previous state can be designed in such a way that the particles of Alice and Thomas have never interacted before. If Bob and Victor perform a Bell state measurement (BSM), it turns out that for any of the outcomes the particles of Alice and Thomas will collapse to some Bell state. By exploiting classical communication Bob and Victor can send Thomas the measurement outcomes, and then Thomas can perform local rotations in order to obtain the entangled state $|\phi^+\rangle_{AT}$. After the BSM the particles of Alice and Thomas become entangled although they have never interacted directly before as they can be created by different sources in highly separated places. One sees that this protocol is basically an extension of the teleportation one, where one member of the first Einstein–Podolsky–Rosen (EPR) pair (between Alice and Bob) is teleported to the second EPR pair (between Victor and Thomas). We must keep in mind that any pair can be chosen as the teleported pair or the channel. The idea of entanglement swapping was developed in order to distribute the entanglement over long distances, this is a fundamental feature to implement a quantum repeater [43]. This idea was also generalized to multipartite scenarios [44] which are particularly useful when working with quantum cryptography.

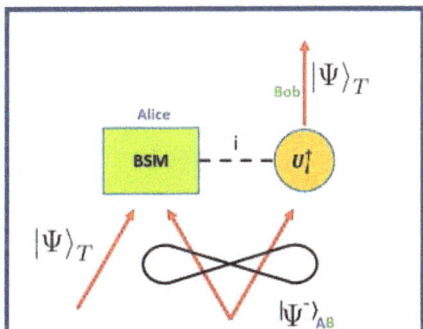

Figure 1. Pictorial representation of a teleportation protocol. Description of a teleportation protocol. The two stations A and B share an entangled state and a classical communication channel (dashed black line), which is used to communicate the result i of the Bell state measurement (BSM) performed in A in order to drive a unitary operation U_i. The initial quantum state $|\Psi\rangle_T$, which is physically present in A, is thus teleported in B.

3. The First Teleportation

Following the original teleportation paper and its continuous-variables version, an intensive experimental effort started for the experimental realization of teleportation. Here we focus on the first experiment carried out in Rome. As recently reported by Nicolas Gisin in Nature [45]:

> "Two groups achieved the feat of quantum teleportation in 1997—just four years after the theoretical breakthrough. First, it was the team of Boschi et al. based in Italy, followed only a few months later by the team of Bouwmeester et al. in Austria."

The scheme adopted in Rome exploited the approach proposed by Sandu Popescu in 1995 [46]. A total of two photons, rather than three as done in Innsbruck by Zeilinger's group [16], were used. Let us briefly summarize the description provided in Figure 2. The two photon entangled state exhibited a path entanglement while the polarization degree of freedom of one of the photons was employed for preparing the unknown state. This approach avoided the difficulties associated with having three photons, as done in [16], and made the Bell measurement complete. This scheme is equivalent to the original scheme up to a local operation (since, in principle, any unknown state of a particle from outside could be swapped onto the polarization degree of freedom of Alice's EPR particle by a local unitary operation as discussed below). In particular, if the preparer does not tell Alice what state he has prepared then there is no way Alice can find out what the state is. It is worthwhile mentioning that this approach leads to a 100% success rate for the Bell measurement in the ideal case rather than 50% as in three photon based schemes.

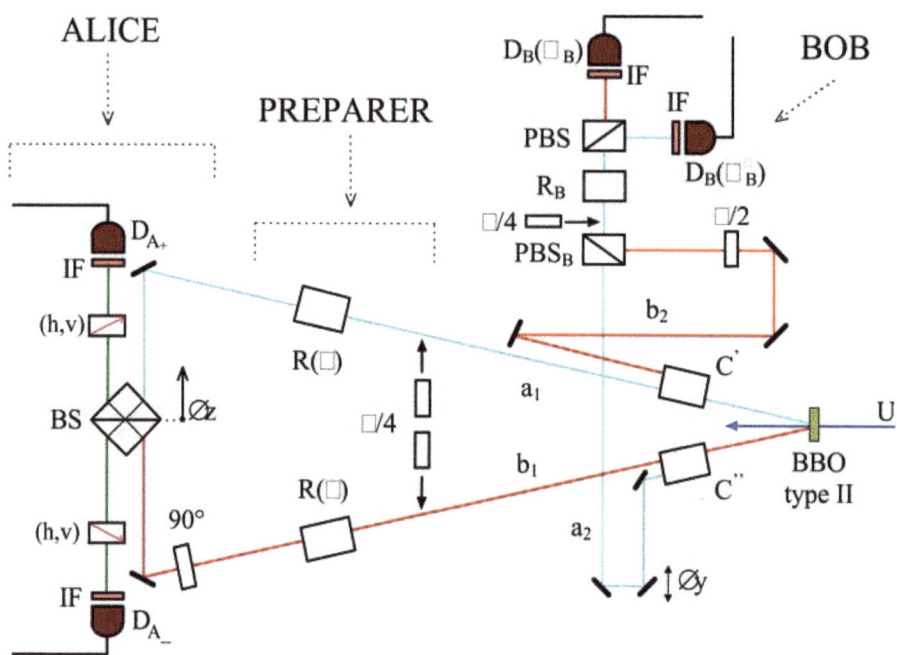

Figure 2. Experimental scheme adopted in the 1998 experiment showing the separate roles of the preparer, Alice and Bob. Pairs of polarization entangled photons are created directly using type II degenerate parametric down-conversion. By means of quarter wave plates acting in the same way on paths a_1 and b_1 the polarization degree of freedom of photon 1 is used to prepare the state to be teleported. For Alice, the polarization of path b_1 is first rotated by a further 90°. Then paths a_1 and b_1 impinge on the two input ports of an ordinary 50:50 beamsplitter (BS). At this beamsplitter each of the two polarizations h, and v interfere independently. After the beamsplitter there are two polarizers which are set either to transmit h or to transmit v polarization to the detector $D_{A_\pm}$. At Bob's end, path b_2 is rotated through 90° by a half waveplate. The paths a_2 and b_2 are combined at a polarizing beamsplitter orientated to transmit vertical and reflect horizontal polarization, then letting it impinge on a polarizing beamsplitter followed by two detectors $D_B(\theta_B)$. PBS, IF, and BBO stand, respectively, for Polarizing Beam Splitter, Interferential Filter and Beta Barium Borate. Picture from [17].

The scheme adopted in the experimental realization is reported in Figure 2. Pairs of polarization entangled photons were created directly using type-II degenerate parametric down-conversion. The article reported results for the teleportation of a linearly polarized state and of an elliptically polarized state. It showed that the experimental results cannot be explained in terms of a classical channel alone. The Bell measurement could distinguish between all four Bell states simultaneously allowing, in the ideal case, a 100% success rate of teleportation. As said, this scheme exploited two degrees of freedom of the same particle (polarization and path) to implement two different qubits:

This approach allows the achievement of a deterministic Control-NOT gate leading to a complete Bell state measurement apparatus.

Let us note that the merging of the polarization quantum state of two photons into one photon has been recently reported by the Roma group in collaboration with the University of Naples Federico II [47]. This physical process has been named 'quantum joining', in which the two-dimensional quantum states of two input photons are combined into a single output photon, within a four-dimensional Hilbert space. This process provides a flexible quantum interconnect to bridge multi-particle protocols of quantum information with multidegree-of-freedom ones. Hence, by exploiting the quantum joining, it is possible to join the quantum state to be teleported with the photon A of the entangled pair. By this approach it is then possible to teleport any external quantum states via the "Roma" teleportation scheme. The scheme demonstrated in [47] is probabilistic with a success probability equal to 1/8, to be compared with the success probability of 1/2 for the scheme adopted by [16]. Nevertheless it is possible to enhance the merging probability up to 1 by increasing the number of ancillary photons and then the complexity of the related scheme [48]. Alternatively, by adopting gigantic nonlinear interactions among photons currently under development [49], deterministic schemes for quantum-state joining and splitting should also become possible [48].

The experiment carried out in Rome was submitted to Physical Review Letters on 28 July 1997 and posted on arXiv 2 October 1997 [17]. We refer to [50] for a complete comparison between the experiments carried out in Rome [17] and Innsbruck [16]. Here we are not describing how to adapt the QST protocol to continuous variable systems [26]: a very exhaustive review on these concept and implementations can be found in [12]. We refer to [51] and [12] for an exhaustive description of Quantum Teleportation with Continuous Variables. It is worth mentioning that a long debate has addressed the differences between unconditional (or deterministic) and conditional quantum teleportation: the different points of view, respectively, of the continuous and discrete variable community are properly summarized in [12] and [52].

4. Teleportation of Vacuum-One Photon

The Roma team addressed a qubit teleportation with a large fidelity by adopting the concept of entanglement of one photon with the vacuum [18]. The underlying motivation was to identify and implement the simplest scheme to observe the essence of the teleportation of a quantum state. By this approach, each quantum superposition state, i.e., a qubit, was physically implemented by a two dimensional subspace of Fock states of a mode of the electromagnetic field, specifically the space spanned by the "vacuum" and the 1-photon state. In other words, the field's modes rather than the photons associated with them have been properly taken as the information and entanglement carriers.

The following details are taken from reference [53], where a complete description of the scheme and related experiment is available. If A and B represent two different modes of the field, with wavevectors k_A and k_B directed respectively towards two distant stations (Alice and Bob), these ones may be linked by a non-local channel expressed by an entangled state implying the quantum superposition of a single photon, e.g., by the singlet: $|\Psi^-\rangle_{A,B} = \frac{1}{\sqrt{2}}(|0\rangle_{kA}|1\rangle_{kB} - |1\rangle_{kA}|0\rangle_{kB})$ Here the mode indexes 0 and 1 denote respectively the vacuum and 1-photon Fock state population of the modes k_A, k_B and the superposition state may be simply provided by an optical beam splitter (BS), as we shall see.

Conceptually this experiment represents one of the first (if not the first) application of "single-photon nonlocality", a paradigm first introduced by Albert Einstein in a context close to the formulation of the Einstein-Podolsky-Rosen paradox [54] and later elaborated by [55,56]. Moreover this scheme is highly connected with single particle entanglement adopted as a key resource in the method proposed by Knill, Laflamme and Milburn [57] to implement universal quantum computing with linear optics.

Of course, in order to make use of the entanglement present in this picture we need to use the second quantization procedure of creation and annihilation of particles and/or use states which are superpositions of states with different numbers of particles. Another puzzling aspect of this second quantized picture is the need to define and measure the relative phase between states with different number of photons, such as the relative phase between the vacuum and one photon state. In order to control these relative phases we need, in analogy with classical computers, to supply all gates and all sender/receiving stations of a quantum information network with a common clock signal, e.g., provided by an ancillary photon or by a multi-photon, Fourier transformed coherent pulse.

The quantum system whose state we want to teleport is a qubit defined on the Hilbert space spanned by the vacuum state $|0\rangle_S$ and the one Fock-state $|1\rangle_S$ of the mode k_S. Thus the mode k_S can be considered the qubit to be teleported. Suppose now that the qubit k_S is in an arbitrary pure state $\alpha |0\rangle_S + \beta |1\rangle_S$. The overall state of the system and the non-local channel is then:

$$
\begin{aligned}
|\Phi_{total}\rangle &= 2^{-\frac{1}{2}} (\alpha |0\rangle_S + \beta |1\rangle_S) (|1\rangle_A |0\rangle_B - |0\rangle_A |1\rangle_B) \\
&= 2^{-\frac{1}{2}} \alpha |\Psi^1\rangle_{SA} |1\rangle_B + 2^{-\frac{1}{2}} \alpha |\Psi^2\rangle_{SA} |0\rangle_B + \\
&\quad 2^{-1} |\Psi^3\rangle_{SA} (\alpha |0\rangle_B + \beta |1\rangle_B) + \\
&\quad 2^{-1} |\Psi^4\rangle_{SA} (\alpha |0\rangle_B - \beta |1\rangle_B)
\end{aligned}
\qquad (6)
$$

where the states $\left|\Psi^j_{SA}\right\rangle$, $j = (1,2,3,4)$ are defined below in Equations (7)–(10). The teleportation proceeds with Alice performing a partial Bell measurement. She combines the modes k_S and k_A on a symmetric beam splitter BS_A whose output modes k_1 and k_2 are coupled to two detectors D_1 and D_2, respectively. As a consequence, we obtain

$$
\begin{aligned}
\left|\Psi^1_{SA}\right\rangle &= |0\rangle_S |0\rangle_A = |0\rangle_1 |0\rangle_2 & (7) \\
\left|\Psi^2_{SA}\right\rangle &= |1\rangle_S |1\rangle_A = 2^{-\frac{1}{2}} (|2\rangle_1 |0\rangle_2 - |0\rangle_1 |2\rangle_2) & (8) \\
\left|\Psi^3_{SA}\right\rangle &= 2^{-\frac{1}{2}} (|0\rangle_S |1\rangle_A - |1\rangle_S |0\rangle_A) = |1\rangle_1 |0\rangle_2 & (9) \\
\left|\Psi^4_{SA}\right\rangle &= 2^{-\frac{1}{2}} (|0\rangle_S |1\rangle_A + |1\rangle_S |0\rangle_A) = |0\rangle_1 |1\rangle_2. & (10)
\end{aligned}
$$

The state $|\Psi^3_{SA}\rangle$ is a Bell type state. From Equation (9) we see that its realization implies a single photon arriving at the detector D_1 and no photons at D_2 in Figure 3. Similarly, $|\Psi^4_{SA}\rangle$ is a Bell type state and it implies a single photon arriving at the detector D_2 and no photons at D_1. In both these cases the teleportation is successful. On the other hand, when Alice finds $|\Psi^1_{SA}\rangle$ or $|\Psi^2_{SA}\rangle$ the teleportation fails. From Equation (6) we see that teleportation is successful in 50% of the cases. By using appropriate entangled resources the teleportation step can be made near deterministic by means of linear optics, photon counting and fast feedforward.

5. Active Teleportation

Up to 2002 all the implementation of quantum state teleportation, including the one reported in the previous Sections, corresponded to simplified "passive" schemes where the transformation U_i at Bob's side was not implemented. In all these experiments the success of the protocol was demonstrated indirectly by the detection of the correlations established a posteriori between the extreme stations, Alice and Bob. These passive realizations had the advantage of avoiding the difficult implementation of the final stage of the protocol, i.e., of the unitary transformations U_i restoring the exact input qubit at Bob's site depending on the outcome of Alice's Bell measurement. The main problem faced here was due to the relatively long time needed to activate, under single-photon excitation by Alice's Bell-measurement apparatus, an Electro-Optic Pockels cell, which implements the necessary U-unitaries at Bob's site. The following details are again taken from [53], where a complete description is available.

The work realized in [19] reported for the first time the complete, i.e., active, qubit teleportation process by completing the corresponding optical scheme according to the full quantum teleportation protocol. This achievement was accomplished exploiting the concept of vacuum-one photon qubit introduced in the previous section. The experimental setup, Figure 4, can be somewhat considered to be the "folded" configuration of the one reported in Figure 3. The significant changes consisted of the addition of the optical delay line and of a different measurement apparatus at Bob's site.

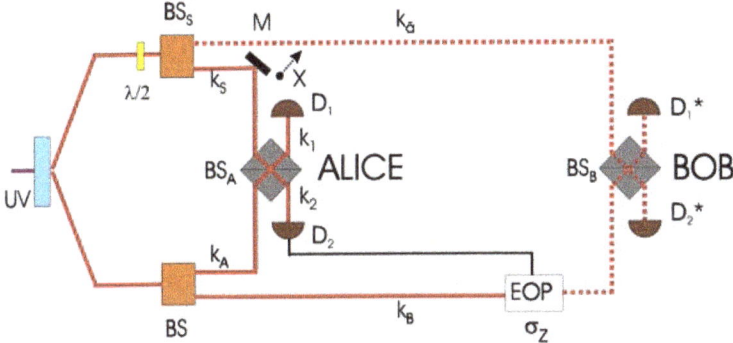

Figure 3. Experimental scheme adopted in the 2002 teleportation of a vacuum–one-photon qubit. EOP denotes a high-voltage Electro-Optic Pockels cell, BS denote beam splitter and D detectors. Picture from [18].

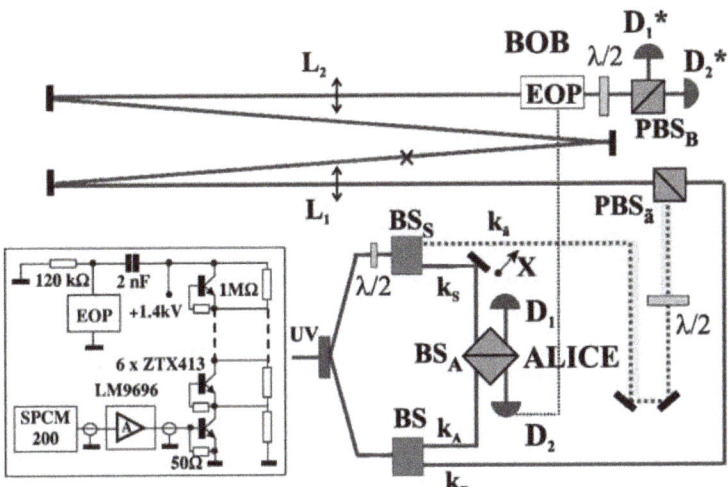

Figure 4. "Active" teleportation of a quantum bit. The experimental set-up can be somewhat considered to be the "folded" configuration of the one reported in Figure 2. The significant changes consisted of the addition of the optical delay line (DL) and of a different measurement apparatus at Bob's site where a high-voltage micro Electro Optics Pockels cell (EOP) performs a unitary transformation $U \equiv \sigma_z$. In the inset is reported the diagram of the fast electronic switch of (EOP). Picture from [19].

6. Optimal Quantum Machines Based on Teleportation

In 2002, the focus of the Rome research activities moved from the implementation of quantum state teleportation protocol to the physical realization of different optimal quantum machines. Let us first briefly summarize the scientific background. We will then highlight the connection with the QST.

At a fundamental level quantum information (QI) consists of the set of rules that identify and characterize the physical transformations that are applicable to the quantum state of any information system. Because of the constraints established by the quantum rules it is found that several classical information tasks are forbidden or cannot be perfectly extended to the quantum world. A well known and relevant QI limitation consists of the impossibility of perfectly cloning (copying) any unknown qubit $|\phi\rangle$ [58]. In other words, the map $|\phi\rangle \rightarrow |\phi\rangle |\phi\rangle$ cannot be realized by nature because it does not belong to the set of completely positive (CP) maps. Another forbidden operation is the NOT gate that maps any $|\phi\rangle$ in its orthogonal state $|\phi^\perp\rangle$ [59]. Even if these two processes are unrealizable in their exact forms, they can be optimally approximated by the so-called universal optimal quantum machines, which exhibit the minimum possible noise.

A complete understanding of these processes is important since the exact characterization of the quantum constraints within basically simple QI processes is useful to design more sophisticated algorithms and protocols and to assess the limit performance of complex networks. The efficiency of a gate, that measures how close its action is to the desired one, is generally quantified by the fidelity F. $F = 1$ implies a perfect implementation, while noisy processes correspond to: $F < 1$. The universal NOT (UNOT) gate, the optimal approximation of the NOT gate, maps N identical input qubits $|\phi\rangle$ into M optimal flipped ones in the state σ_{out}. It achieves the fidelity: $F^*_{N \rightarrow M}(|\phi^\perp\rangle, \sigma_{out}) = \langle \phi^\perp | \sigma_{out} | \phi^\perp \rangle = (N+1)/(N+2)$ that depends only on the number of the input qubits [60]. Indeed the fidelity of the UNOT gate is exactly the same as the optimal quantum estimation fidelity [61]. This means that such process may be modeled as a "classical", i.e., exact, preparation of M identical flipped qubits following the quantum, i.e., inexact, estimation of N input states. Only this last operation is affected by noise. Differently from the UNOT gate, the universal optimal quantum cloning machine (UOQCM), which transforms N identical qubits $|\phi\rangle$ into M identical copies ρ_{out}, achieves as optimal fidelity: $F_{N \rightarrow M}(|\phi\rangle, \rho_{out}) = \langle \phi | \rho_{out} | \phi \rangle = (N + 1 + \beta)/(N + 2)$ with $\beta = N/M \leq 1$ [62–64]. As we can see $F_{N \rightarrow M}(|0\rangle, \rho_{out})$ is larger than the one obtained by the N estimation approach and reduces to that result for $\beta \rightarrow 0$, i.e., for an infinite number of copies. The extra positive term β in the above expression accounts for the excess of quantum information which is originally stored in N states and is optimally redistributed by entanglement among the $M - N$ remaining blank qubits encoded by UOQCM. The UNOT gate and the UOQCM can be implemented following two different approaches:

(i) The first one has been based on finding a suitable unitary operator U_{NM}, acting on N input qubits and on $2(M - N)$ ancillary qubits: Figure 5a. At the output of this device we obtain M and $M - N$ qubits which are, respectively, the optimal clones and the best flipped qubits of the input ones. The transformation U_{NM} can be deterministically realized by means of a quantum network, as proposed by Buzek et al. [65].

(ii) The second approach to implement the $N \rightarrow M$ cloning and the $N \rightarrow (M - N)$ flipping is a probabilistic method that exploits a symmetrization process: Figure 5b. The initial state of the overall system consists of the N input qubits and of $(M - N)$ pairs of entangled qubits. The two optimal quantum machines are performed by applying a projective operation on the symmetric subspace to the N input qubits and to $(M - N)$ ancilla qubits, each one belonging to a different entangled pair. This scheme corresponds to a modified QST scheme: Instead of performing a Bell state measurement a project over the symmetric subspace is performed. This transformation assures the uniform distribution of the initial information into the overall system and guarantees that all output clone qubits are indistinguishable. The success probability is equal to $\frac{1}{2^{M-N}} \frac{1+M}{1+N}$. The $(M - N)$ optimal flipped qubits are teleported in a different location since there is no interaction between the N input qubits and the $(M - N)$ flipped ones.

Qubits Symmetrization: Linear Optics Implementation

Let us consider the scenario where there is $N=1$ initial qubit and the goal is to obtain $M=2$ optimal clones and 1 optimal flipped qubit.

The protocol that realizes the $1 \to 2$ UOQCM and $1 \to 1$ Tele-UNOT gate, involves two distant partners: Alice (A) and Bob (B). A holds the unknown input qubit S in a generic state $|\phi\rangle_S$, while B shall finally receive this qubit encoded optimally by the UNOT transformation of $|\phi\rangle_S$. Let A and B share the entangled singlet state of two qubits A,B: $|\Psi^-\rangle_{AB} = 2^{1/2}(|\phi\rangle_A|\phi^\perp\rangle_B - |\phi^\perp\rangle_A|\phi\rangle_B)$, as in a quantum teleportation protocol [1]. The choice of the singlet state guarantees, in virtue of its SU(2) invariance, the universality of the overall process. The overall state of the system reads $|\Omega\rangle_{SAB} = 2^{-1/2}|\phi\rangle_S(|\phi\rangle_A|\phi^\perp\rangle_B - |\phi^\perp\rangle_A|\phi\rangle_B)$. Let A to apply to the overall initial state $|\Omega\rangle_{SAB}$ the projective operator P_{SA} over the symmetric subspace of the qubits S and A:

$$P_{SA} = (\mathbf{I}_{SA} - |\Psi^-\rangle_{SA}\langle\Psi^-|_{SA}). \tag{11}$$

The projection is successfully realized with probability $p = 3/4$. In this case the normalized output state is:

$$|\Theta\rangle_{SAB} = \sqrt{2/3}|\phi\rangle_S|\phi\rangle_A|\phi^\perp\rangle_B \\ -\frac{1}{\sqrt{6}}(|\phi\rangle_S|\phi^\perp\rangle_A + |\phi^\perp\rangle_S|\phi\rangle_A)|\phi\rangle_B. \tag{12}$$

One bit of classical communication sent by A announces to B the success of the symmetrization protocol. Note that the presence of the entangled state $|\Psi^-\rangle_{AB}$ is not strictly necessary for the sole implementation of the quantum cloning process. Indeed, for this purpose, we could apply P_{SA} to the initial state $|\phi\rangle_S\langle\phi|_S \otimes \frac{I_A}{2}$ as shown in [66,67].

In the experiments reported in [68] and [67], the input qubit was codified into the polarization state of a single photon belonging to the input mode k_S : $|\phi\rangle_S = \alpha|H\rangle_S + \beta|V\rangle_S$, whereas an entangled pair $|\Psi^-\rangle_{AB}$ of photons A and B, was generated on the modes k_A and k_B by spontaneous parametric down conversion (SPDC). The projective operation in the space $H = H_A \otimes H_S$ was realized exploiting the linear superposition of the modes k_S and k_A generated by a 50:50 beam-splitter, BS_A (Figure 6). This superposition allows a partial Bell measurement on the BS_A output states which is needed to implement the cloning machine and the Tele-UNOT gate. Consider the overall output state realized on the two modes k_1 and k_2 of the BS_A and expressed by a superposition of the Bell states: $(|\Psi^-\rangle_{SA}, |\Psi^+\rangle_{SA}, |\Phi^-\rangle_{SA}, |\Phi^-\rangle_{SA})$. The realization of the singlet $|\Psi^-\rangle_{SA}$ is identified by the emission of one photon on each output mode of BS_A, while the realization of the other three Bell states implies the emission of two photons either on mode k_1 or on mode k_2. This Hong-Ou-Mandel interference process, expressing a Bose mode coalescence (BMC) of the two photons over the same mode, was experimentally identified by a coincidence event between two detectors coupled to the output mode k_2 by means of an additional 50:50 beam-splitter by a post-selection technique. The projection into the symmetric space lies at the core of the cloning process.

By this approach two relevant quantum information processes, forbidden by quantum mechanics in their exact form, have been found to be connected contextually by a modified quantum state teleportation scheme in an optimal way. The complete implementation of this protocol has been successfully performed by a fully linear optical setup, which has also been shown to be scalable to a larger number of particles.

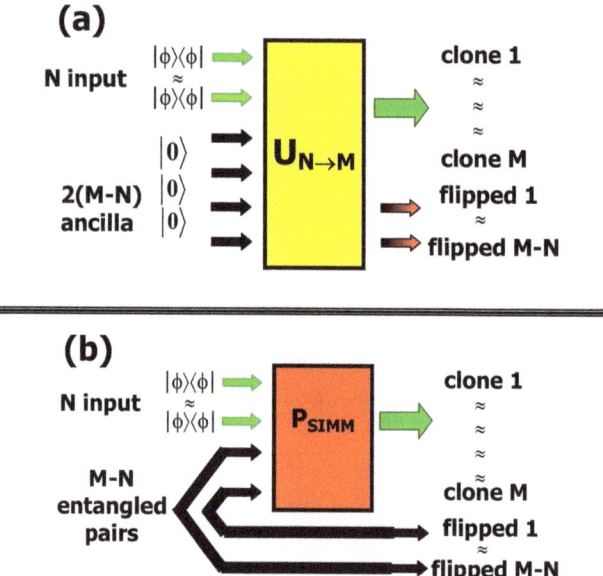

Figure 5. General scheme for the simultaneous realization of the universal NOT (UNOT) gate and of the universal quantum cloning machine. (**a**) Unitary transformation acting on the N-input qubits and 2(M-N) ancilla qubits initially in the state $|0\rangle$. (**b**) Symmetrization process acting on the input qubits and (M-N) entangled pairs of qubits. Picture partially from [68].

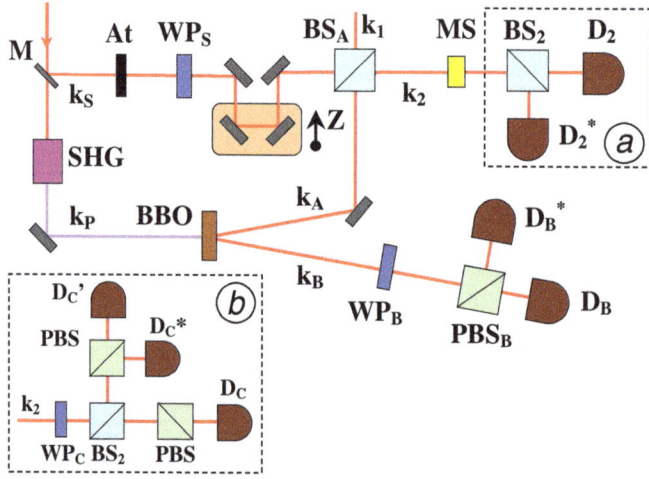

Figure 6. Setup for the optical implementation of the Tele-UNOT Gate and the probabilist universal optimal quantum cloning machine (UOQCM). The measurement setups used for the verification of the cloning and Universal NOT gate experiments are reported, respectively, in the inset (**b**) and (**a**). At, WP, MS, SHG, M, PBS denote, respectively, Attenuation Filter, Wave Plate, Mode Selector, Second Harmonic Generation, Partially Reflecting Mirror, Polarizing Beam Splitter. Picture from [68].

7. Micro and Macro Entanglement

As a following step, the goal has been to extend the previous results to a larger number of particles. To this scope non-linear optics interaction have been exploited for an extended research focused on the theoretical and experimental realization of a macroscopic quantum superposition (MQS) made up of photons. This intriguing, fundamental quantum condition is at the core of the famous argument conceived by Schrodinger in 1935. One of the main experimental challenges to the actual realization of this object resides in unavoidable interactions with the environment, leading to the cancellation of any evidence of the quantum features associated with the macroscopic system.

The experimental scheme adopted a nonlinear process, "quantum-injected optical parametric amplification", which, by a linearized cloning process maps the quantum coherence of a single particle state, i.e., a microqubit, onto a macroqubit consisting of a large number M of photons in quantum superposition: Figure 7. Since the adopted scheme was found resilient to decoherence, a MQS demonstration was carried out experimentally at room temperature with $M = 10^4$. The result led to an extended study of quantum cloning, quantum amplification, and quantum decoherence. Several experiments have been carried out, such as the test of the "nosignaling theorem". In addition, the consideration of the microqubit-macroqubit entanglement regime has been extended to macroqubit-macroqubit conditions. The MQS interference patterns for large M are revealed in the experiment and bipartite microqubit-macroqubit entanglement was also demonstrated for a limited number of generated particles. For a complete description of this activity the reader can refer to [69].

8. Summary and Perspectives

The original work by Bennett et al. [1] has rapidly triggered a large number of investigations [10] for a broad range of applications [12]. Teleportation schemes were shown to enable new approaches for universal quantum computation [15,70], in particular as one-way quantum computers [71]. From the experimental perspective, numerous achievements have been reported on photonic platforms proving the feasibility of the scheme already with state-of-the-art technology. After the first demonstrations in 1997–1998 [16,17], one further proof appeared with the unconditional teleportation of optical coherent states with squeezed-state entanglement [26]. Later on, [72] provided a proof of the nonlocality of the process and of entanglement swapping. One year later, as shown in the previous sections, [18] teleported qubits were encoded in vacuum–one-photon states. The new century witnessed a worldwide race towards more complex implementations. In 2004, a single-mode discrete teleportation scheme using a quantum dot single-photon source has been demonstrated [73] based on the scheme of [18]. At the same time, several experiments addressed teleportation over larger distances [20–22,39]. Teleportation was also reported on squeezed entangled states: We refer to the review [12]. To bridge the gap between discrete and continuous variables, a hybrid approach has been recently reported [74]. Finally, achievements on photonic teleportation have been demonstrated with the first implementation on integrated circuits [75], as well as schemes with simultaneous teleportation of multiple degrees of freedom [27] and teleportation of qudits [76].

The past decade has seen a strong effort directed towards the development of matter-light interfaces as building blocks for quantum computation and communication, where entanglement between single-photon states and atomic ensembles represents an effective solution. The last few years have seen the implementation of quantum teleportation in scenarios of growing complexity. The following step is to exploit these results in order to achieve quantum networks over large distances thanks to the adoption of quantum repeaters. Concerning the research effort, the Quantum Information Lab is currently focused on experimental quantum causality. A promising direction is to exploit teleportation and entanglement swapping within such a framework [77].

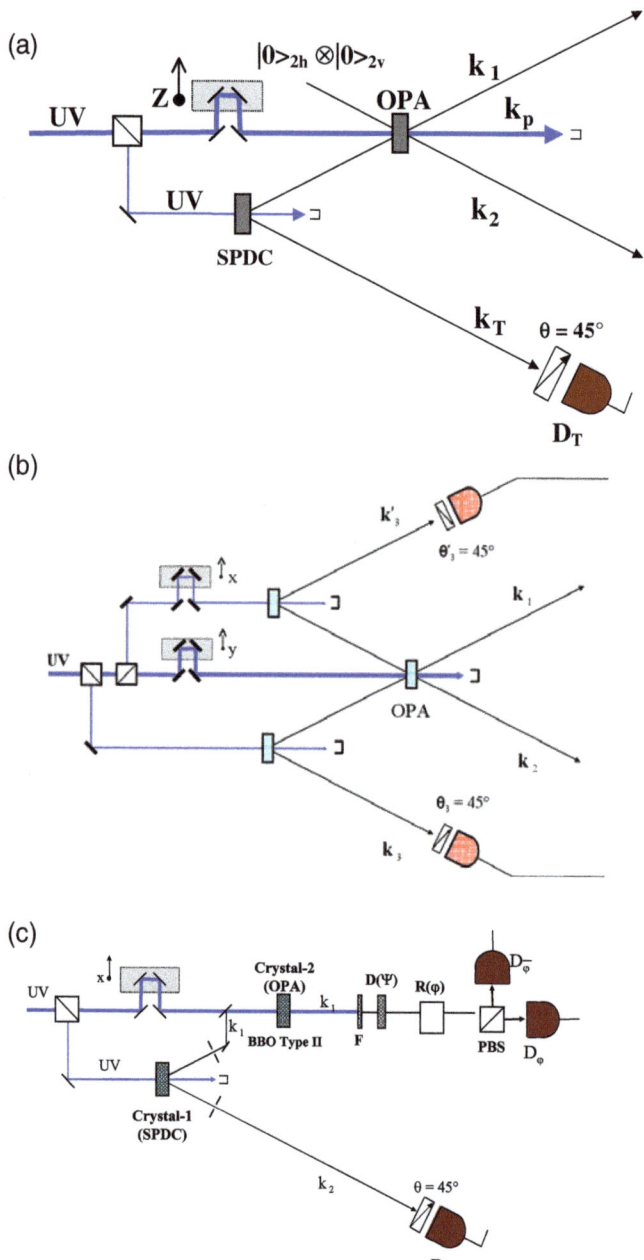

Figure 7. Three different configurations for the amplification of quantum states. (**a**) Schematic diagram of a noncollinear quantum-injected optical parametric amplifier (OPA). The injection is provided by an external spontaneous parametric downconversion source of polarization-entangled photon states. (**b**) Double injection of the optical parametric amplifier. (**c**) Collinear quantum-injected optical parametric amplifier. Picture from [69]. Spontaneous parametric down conversion (SPDC).

Author Contributions: F.D.M. and F.S. equally contributed to the manuscript preparation and writing.

Funding: This research received no external funding.

Conflicts of Interest: The authors declare no conflict of interest.

References

1. Bennett, C.H.; Brassard, G.; Crépeau, C.; Jozsa, R.; Peres, A.; Wootters, W.K. Teleporting an unknown quantum state via dual classical and Einstein-Podolsky-Rosen channels. *Phys. Rev. Lett.* **1993**, *70*, 1895. [CrossRef]
2. Bennett, C.H.; Wiesner, S.J. Communication via one- and two-particle operators on Einstein-Podolsky-Rosen states. *Phys. Rev. Lett.* **1992**, *69*, 2881–2884. [CrossRef]
3. Feynman, R.P. Simulating physics with computers. *Int. J. Theor. Phys.* **1982**, *21*, 467–488. [CrossRef]
4. Shor, P.W. Scheme for reducing decoherence in quantum computer memory. *Phys. Rev. A* **1995**, *52*, R2493–R2496. [CrossRef]
5. Steane, A.M. Error correcting codes in quantum theory. *Phys. Rev. Lett.* **1996**, *77*, 793–797. [CrossRef]
6. Ekert, A.K. Quantum cryptography based on Bell's theorem. *Phys. Rev. Lett.* **1991**, *67*, 661–663. [CrossRef]
7. Jennewein, T.; Simon, C.; Weihs, G.; Weinfurter, H.; Zeilinger, A. Quantum cryptography with entangled photons. *Phys. Rev. Lett.* **2000**, *84*, 4729–4732. [CrossRef]
8. Acin, A.; Brunner, N.; Gisin, N.; Massar, S.; Pironio, S.; Scarani, V. Device-Independent security of quantum cryptography against collective attacks. *Phys. Rev. Lett.* **2007**, *98*, 230501. [CrossRef]
9. Brunner, N.; Cavalcanti, D.; Pironio, S.; Scarani, V.; Wehner, S. Bell nonlocality. *Rev. Mod. Phys.* **2014**, *86*, 419. [CrossRef]
10. Nielsen, M.A.; Chuang, I.L. *Quantum Computation and Quantum Information*; Cambridge University Press: Cambridge, UK, 2000.
11. Weedbrook, C.; Pirandola, S.; García-Patrón, R.; Cerf, N.J.; Ralph, T.C.; Shapiro, J.H.; Lloyd, S. Gaussian Quantum Information. *Rev. Mod. Phys.* **2012**, *84*, 621. [CrossRef]
12. Pirandola, S.; Eisert, J.; Weedbrook, C.; Furusawa, A.; Braunstein, S.L. Advances in quantum teleportation. *Nature Photon.* **2015**, *9*, 641–652. [CrossRef]
13. Briegel, H.-J.; Dur, W.; Cirac, J.I.; Zoller, P. Quantum repeaters: The role of imperfect local operations in quantum communication. *Phys. Rev. Lett.* **1998**, *81*, 5932. [CrossRef]
14. Dur, W.; Briegel, H.-J.; Cirac, J.I.; Zoller, P. Quantum repeaters based on entanglement purification. *Phys. Rev. A* **1999**, *60*, 725. [CrossRef]
15. Gottesman, D.; Chuang, I.L. Demonstrating the viability of universal quantum computation using teleportation and single-qubit operations. *Nature* **1999**, *402*, 6390–6393. [CrossRef]
16. Bouwmeester, D.; Pan, J.-W.; Mattle, K.; Eibl, M.; Weinfurter, H.; Zeilinger, A. Experimental quantum teleportation. *Nature* **1997**, *390*, 575–579. [CrossRef]
17. Boschi, D.; Branca, S.; De Martini, F.; Hardy, L.; Popescu, S. Experimental realisation of teleporting an unknown pure quantum state via dual classical and Einstein–Podolski–Rosen channels. *Phys. Rev. Lett.* **1998**, *80*, 1121–1125. [CrossRef]
18. Lombardi, E.; Sciarrino, F.; Popescu, S.; De Martini, F. Teleportation of a Vacuum–One-Photon Qubit. *Phys. Rev. Lett.* **2002**, *88*, 070402. [CrossRef]
19. Giacomini, S.; Sciarrino, F.; Lombardi, E.; De Martini, F. Active teleportation of a quantum bit. *Phys. Rev. A* **2002**, *66*, 030302. [CrossRef]
20. Ursin, R.; Jennewein, T.; Aspelmeyer, M.; Kaltenbaek, R.; Lindenthal, M.; Walther, P.; Zeilinger, A. Quantum teleportation across the Danube. *Nature* **2004**, *430*, 849. [CrossRef]
21. Marcikic, I.; de Riedmatten, H.; Tittel, W.; Zbinden, H.; Gisin, N. Long-distance teleportation of qubits at telecommunication wavelengths. *Nature* **2003**, *421*, 509–513. [CrossRef]
22. De Riedmatten, H.; Marcikic, I.; Tittel, W.; Zbinden, H.; Collins, D.; Gisin, N. Long-distance quantum teleportation in a quantum relay configuration. *Phys. Rev. Lett.* **2004**, *92*, 047904. [CrossRef]
23. Yin, J.; Ren, J.G.; Lu, H.; Cao, Y.; Yong, H.L.; Wu, Y.P.; Liu, C.; Liao, S.K.; Zhou, F.; Jiang, Y.; et al. Quantum teleportation and entanglement distribution over 100-kilometre free-space channels. *Nature* **2012**, *488*, 185–188. [CrossRef]

24. Ma, X.S.; Herbst, T.; Scheidl, T.; Wang, D.; Kropatschek, S.; Naylor, W.; Wittmann, B.; Mech, A.; Kofler, J.; Anisimova, E.; et al. Quantum teleportation over 143 kilometres using active feed-forward. *Nature* **2012**, *489*, 269–273. [CrossRef]
25. Ren, J.G.; Xu, P.; Yong, H.L.; Zhang, L.; Liao, S.K.; Yin, J.; Liu, W.Y.; Cai, W.Q.; Yang, M.; Li, L.; et al. Ground-to-satellite quantum teleportation. *Nature* **2017**, *549*, 70–73. [CrossRef]
26. Furusawa, A.; Sørensen, J.L.; Braunstein, S.L.; Fuchs, C.A.; Kimble, H.J.; Polzik, E.S. Unconditional Quantum Teleportation. *Science* **1998**, *282*, 706. [CrossRef]
27. Wang, X.L.; Cai, X.D.; Su, Z.E.; Chen, M.C.; Wu, D.; Li, L.; Liu, N.L.; Lu, C.Y.; Pan, J.W. Quantum teleportation of multiple degrees of freedom in a single photon. *Nature* **2015**, *518*, 516–519. [CrossRef]
28. Nielsen, M.A.; Knill, E.; Laflamme, R. Complete quantum teleportation using nuclear magnetic resonance. *Nature* **1998**, *396*, 52. [CrossRef]
29. Sherson, J.F.; Krauter, H.; Olsson, R.K.; Julsgaard, B.; Hammerer, K.; Cirac, I.; Polzik, E.S. Quantum teleportation between light and matter. *Nature* **2006**, *443*, 557–560. [CrossRef]
30. Krauter, H.; Salart, D.; Muschik, C.A.; Petersen, J.M.; Shen, H.; Fernholz, T.; Polzik, E.S. Deterministic quantum teleportation between distant atomic objects. *Nat. Phys.* **2013**, *9*, 400–404. [CrossRef]
31. Barrett, M.D.; Chiaverini, J.; Schaetz, T.; Britton, J.; Itano, W.M.; Jost, J.D.; Knill, E.; Langer, C.; Leibfried, D.; Ozeri, R.; et al. Deterministic quantum teleportation of atomic qubits. *Nature* **2004**, *429*, 737–739. [CrossRef]
32. Riebe, M.; Häffner, H.; Roos, C.F.; Hänsel, W.; Benhelm, J.; Lancaster, G.P.T.; Körber, T.W.; Becher, C.; Schmidt-Kaler, F.; James, D.F.V.; et al. Deterministic quantum teleportation with atoms. *Nature* **2004**, *429*, 734–737. [CrossRef] [PubMed]
33. Gao, W.B.; Fallahi, P.; Togan, E.; Delteil, A.; Chin, Y.S.; Miguel-Sanchez, J.; Imamoğlu, A. Quantum teleportation from a propagating photon to a solid-state spin qubit. *Nat. Commun.* **2013**, *4*, 2744. [CrossRef]
34. Steffen, L.; Salathe, Y.; Oppliger, M.; Kurpiers, P.; Baur, M.; Lang, C.; Eichler, C.; Puebla-Hellmann, G.; Fedorov, A.; Wallraff, A. Deterministic quantum teleportation with feed-forward in a solid state system. *Nature* **2013**, *500*, 319–322. [CrossRef]
35. Pfaff, W.; Hensen, B.J.; Bernien, H.; van Dam, S.B.; Blok, M.S.; Taminiau, T.H.; Tiggelman, M.J.; Schouten, R.N.; Markham, M.; Twitchen, D.J.; et al. Unconditional quantum teleportation between distant solid-state quantum bits. *Science* **2014**, *345*, 532–535. [CrossRef]
36. Yin, J.; Cao, Y.; Li, Y.H.; Liao, S.K.; Zhang, L.; Ren, J.G.; Cai, W.Q.; Liu, W.Y.; Li, B.; Dai, H.; et al. Satellite-based entanglement distribution over 1200 kilometers. *Science* **2017**, *356*, 6343. [CrossRef]
37. Valivarthi, R.; Zhou, Q.; Aguilar, G.H.; Verma, V.B.; Marsili, F.; Shaw, M.D.; Nam, S.W.; Oblak, D.; Tittel, W. Quantum teleportation across a metropolitan fibre network. *Nat. Photonics* **2016**, *10*, 677. [CrossRef]
38. Simon, C. Towards a global quantum network. *Nat. Photonics* **2017**, *11*, 678. [CrossRef]
39. Xia, X.-X.; Sun, Q.-C.; Zhang, Q.; Pan, J.-W. Long distance quantum teleportation. *Quantum Sci. Technol.* **2018**, *3*, 014012. [CrossRef]
40. Hensen, B.; Bernien, H.; Dréau, A.E.; Reiserer, A.; Kalb, N.; Blok, M.S.; Ruitenberg, J.; Vermeulen, R.F.; Schouten, R.N.; Abellán, C.; et al. Loophole-free Bell inequality violation using electron spins separated by 1.3 kilometres. *Nature* **2015**, *526*, 682. [CrossRef]
41. Zukowski, M.; Zeilinger, A.; Horne, M.A.; Ekert, A.K. Event-ready detectors Bell experiment via entanglement swapping. *Phys. Rev. Lett.* **1993**, *71*, 4287–4290. [CrossRef]
42. Schmid, C.; Kiesel, N.; Weber, U.K.; Ursin, R.; Zeilinger, A.; Weinfurter, H. Quantum teleportation and entanglement swapping with linear optics logic gates. *New J. Phys.* **2009**, *11*, 033008. [CrossRef]
43. Sangouard, N.; Simon, C.; de Riedmatten, H.; Gisin, N. Quantum repeaters based on entanglement purification. *Rev. Mod. Phys.* **2011**, *83*, 33. [CrossRef]
44. Bose, S.; Vedral, V.; Knight, P.L. Multiparticle generalization of entanglement swapping. *Phys. Rev. A* **1998**, *57*, 822. [CrossRef]
45. Gisin, N. Quantum-teleportation experiments turn 20. *Nature* **2017**, *552*, 42–43. [CrossRef] [PubMed]
46. Popescu, S. An optical method for teleportation. *arXiv* **1995**, arXiv:quant-ph/9501020.
47. Vitelli, C.; Spagnolo, N.; Aparo, L.; Sciarrino, F.; Santamato, E.; Marrucci, L. Joining the quantum state of two photons into one. *Nat. Photonics* **2013**, *7*, 521. [CrossRef]
48. Passaro, E.; Vitelli, C.; Spagnolo, N.; Sciarrino, F.; Santamato, E.; Marrucci, L. Joining and splitting the quantum states of photons. *Phys. Rev. A* **2013**, *88*, 062321. [CrossRef]

49. Peyronel, T.; Firstenberg, O.; Liang, Q.-Y.; Hofferberth, S.; Gorshkov, A.V.; Pohl, T.; Lukin, M.D.; Vuletic, V. Attractive photons in a quantum nonlinear medium. *Nature* **2012**, *488*, 57. [CrossRef]
50. Sudbery, T. The fastest way from A to B. *Nature* **1997**, *390*, 551–552. [CrossRef]
51. Pirandola, S.; and Mancini, S. Quantum Teleportation with Continuous Variables: a survey. *Laser Phys.* **2006**, *16*, 418. [CrossRef]
52. Pan, J.-W.; Chen, Z.B.; Lu, C.-Y.; Weinfurter, H.; Zeilinger, A.; Zukowski, M. Multi-photon entanglement and interferometry. *Rev. Mod. Phys.* **2012**, *84*, 777. [CrossRef]
53. Sciarrino, F.; Lombardi, E.; Giacomini, S.; De Martini, F. Active teleportation and entanglement swapping of a vacuum-one photon qubit. *Fortschr. Phys.* **2003**, *51*, 331–341. [CrossRef]
54. Bohr, N. *Albert Einstein: Philosopher-Scientist*; Schlipp, P.A. Ed.; Northwestern University Press: Evanston, IL, USA, 1949.
55. Tan, S.; Walls, D.; Collet, M. Nonlocality of a single photon. *Phys. Rev. Lett.* **1991**, *66*, 252. [CrossRef] [PubMed]
56. Hardy, L. Nonlocality of a Single Photon Revisited. *Phys. Rev. Lett.* **1994**, *73*, 2279. [CrossRef] [PubMed]
57. Knill, E.; Laflamme, R.; Milburn, G. A scheme for efficient quantum computation with linear optics. *Nature* **2001**, *409*, 46. [CrossRef] [PubMed]
58. Wooters, W.K.; Zurek, W.H. A single quantum cannot be cloned. *Nature* **1982**, *299*, 802. [CrossRef]
59. Bechmann-Pasquinucci, H.; Gisin, N. Incoherent and coherent eavesdropping in the six-state protocol of quantum cryptography. *Phys. Rev. A* **1999**, *59*, 4238. [CrossRef]
60. Buzek, V.; Hillery, M.; Werner, R.F. Optimal manipulations with qubits: Universal-NOT gate. *Phys. Rev. A* **1999**, *60*, 2626. [CrossRef]
61. Massar, S.; Popescu, S. Optimal Extraction of Information from Finite Quantum Ensembles. *Phys. Rev. Lett.* **1995**, *74*, 1259. [CrossRef]
62. Buzek, V.; Hillery, M. Quantum copying: Beyond the no-cloning theorem. *Phys. Rev. A* **1996**, *54*, 1844. [CrossRef]
63. Gisin, N.; Massar, S. Optimal Quantum Cloning Machines. *Phys. Rev. Lett.* **1997**, *79*, 2153. [CrossRef]
64. Buzek, V.; Hillery, M. Universal Optimal Cloning of Arbitrary Quantum States: From Qubits to Quantum Registers. *Phys. Rev. Lett.* **1998**, *81*, 5003. [CrossRef]
65. Buzek, V.; Braunstein, S.L.; Hillery, M.; Bruß, D. Quantum copying: A network. *Phys. Rev. A* **1997**, *56*, 3446. [CrossRef]
66. Werner, R.F. Optimal cloning of pure states. *Phys. Rev. A* **1998**, *58*, 1827. [CrossRef]
67. Sciarrino, F.; Sias, C.; Ricci, M.; De Martini, F. Realization of universal optimal quantum machines by projective operators and stochastic maps. *Phys. Rev. A* **2004**, *70*, 052305. [CrossRef]
68. Ricci, M.; Sciarrino, F.; Sias, C.; De Martini, F. Teleportation Scheme Implementing the Universal Optimal Quantum Cloning Machine and the Universal NOT Gate. *Phys. Rev. Lett.* **2004**, *92*, 047901. [CrossRef]
69. De Martini, F.; Sciarrino, F. Colloquium: Multiparticle quantum superpositions and the quantum-to-classical transition. *Rev. Mod. Phys.* **2012**, *84*, 1766. [CrossRef]
70. Ishizaka, S.; Hiroshima, T. Asymptotic Teleportation Scheme as a Universal Programmable Quantum Processor. *Phys. Rev. Lett.* **2008**, *101*, 240501. [CrossRef]
71. Raussendorf, R.; Briegel, H.J. A One-Way Quantum Computer. *Phys. Rev. Lett.* **2001**, *86*, 5188. [CrossRef]
72. Jennewein, T.; Weihs, G.; Pan, J.-W.; Zeilinger, A. Experimental nonlocality proof of quantum teleportation and entanglement swapping. *Phys. Rev. Lett.* **2001**, *88*, 017903. [CrossRef]
73. Fattal, D.; Diamanti, E.; Inoue, K.; Yamamoto, Y. Quantum Teleportation with a Quantum Dot Single Photon Source. *Phys. Rev. Lett.* **2004**, *92*, 037904. [CrossRef] [PubMed]
74. Takeda, S.; Mizuta, T.; Fuwa, M.; van Loock, P.; Furusawa, A. Deterministic quantum teleportation of photonic quantum bits by a hybrid technique. *Nature* **2013**, *500*, 315. [CrossRef] [PubMed]
75. Metcalf, B.J.; Spring, J.B.; Humphreys, P.C.; Thomas-Peter, N.; Barbieri, M.; Kolthammer, W.S.; Jin, X.M.; Langford, N.K.; Kundys, D.; Gates, J.C.; et al. Quantum teleportation on a photonic chip. *Nat. Photon.* **2014**, *8*, 770. [CrossRef]

76. Goyal, S.K.; Boukama-Dzoussi, P.E.; Ghosh, S.; Roux, F.S.; Konrad, T. Qudit-teleportation for photons with linear optics. *Sci. Rep.* **2014**, *4*, 4543. [CrossRef] [PubMed]
77. Carvacho, G.; Chaves, R.; Sciarrino, F. Perspectives on experimental quantum causality. *Europhys. Lett.* **2019**, *125*, 3. [CrossRef]

© 2019 by the authors. Licensee MDPI, Basel, Switzerland. This article is an open access article distributed under the terms and conditions of the Creative Commons Attribution (CC BY) license (http://creativecommons.org/licenses/by/4.0/).

Article

Entanglement 25 Years after Quantum Teleportation: Testing Joint Measurements in Quantum Networks

Nicolas Gisin

Group of Applied Physics, University of Geneva, 1211 Geneva, Switzerland; nicolas.gisin@unige.ch

Received: 27 September 2018; Accepted: 20 March 2019; Published: 26 March 2019

Abstract: Twenty-five years after the invention of quantum teleportation, the concept of entanglement gained enormous popularity. This is especially nice to those who remember that entanglement was not even taught at universities until the 1990s. Today, entanglement is often presented as a resource, the resource of quantum information science and technology. However, entanglement is exploited twice in quantum teleportation. Firstly, entanglement is the "quantum teleportation channel", i.e., entanglement between distant systems. Second, entanglement appears in the eigenvectors of the joint measurement that Alice, the sender, has to perform jointly on the quantum state to be teleported and her half of the "quantum teleportation channel", i.e., entanglement enabling entirely new kinds of quantum measurements. I emphasize how poorly this second kind of entanglement is understood. In particular, I use quantum networks in which each party connected to several nodes performs a joint measurement to illustrate that the quantumness of such joint measurements remains elusive, escaping today's available tools to detect and quantify it.

Keywords: quantum teleportation; quantum measurements; nonlocality

1. Introduction

In 1993 six co-authors surprised the world by proposing a method to teleport a quantum state from one location to a distant one [1,2]. Twenty five years later the surprise is gone, but the fascination remains; how can an object submitted to the no-cloning theorem disappear here and reappear there without anything carrying any information about it transmitted from the sender, Alice, to the receiver, Bob? Today, the answer seems well known and has a name: entanglement [3]. This merely shifts the mystery, and thus the fascination, to entanglement. However, entanglement appears twice in quantum teleportation. The first and most obvious appearance of entanglement is as the "quantum teleportation channel", i.e., entanglement between two systems, the first one controlled by Alice, the second one controlled by Bob. This sort of entanglement is by now pretty well (though no fully) understood. But entanglement appears a second time in quantum teleportation: the measurement that Alice has to perform jointly on the quantum state to be teleported and her half of the "quantum teleportation channel" has all its eigenstates maximally entangled.

Without this second appearance of entanglement, quantum teleportation would be impossible. This can be understood intuitively as follows [4]. First, observe that two (maximally) entangled systems are characterized by the property that if one asks both of them the same question—i.e., perform the same measurement on each of them, then both systems deliver the same answer (see Endnote [5]—which refers to References [1,2]). Well, for singlets it's just the opposite, they get opposite results instead of identical ones, but that's just a matter of systematically flipping one of the answers. Now, the joint measurement essentially asks to two independent systems the following "strange question": "if I would perform the same measurement on both of you, would you provide the same answer?" This is a question about the relation between the two systems, not a pair of questions to each system whose answers are then combined in some clever way. Indeed, classical systems, including humans, can't answer such unusual joint questions. But quantum systems can. For example, the two systems can answer "yes" and get (maximally) entangled in such a way that whatever identical questions are later asked to them, they'll provide the same answer. Or the answer could be "no" and the two systems get into a different (maximally) entangled state such that their answer to arbitrary but identical questions would always be opposite. As is well-known, in order to terminate the quantum teleportation process, Alice has to communicate (classically) which result she obtained to her "strange question". Then Bob knows whether his system will provide the same answer as had the question been asked to the original system, the one to be teleported, or whether he will receive just the opposite answer. It is important to notice that this classical communication from Alice to Bob carries exactly zero information about the teleported quantum state.

Well, in quantum theory the situation is a bit more complicated, with four possible answers to the joint "strange" measurement and a bit more involved relations between the answer and Bob's system. But the essential is there and it calls for understanding! Physics requires an understanding of such joint measurements of similar quality as our understanding of entanglement between distant systems, i.e., of entanglement as quantum teleportation channels. The quality of today's understanding of entanglement between distant systems is illustrated by its relation to Bell non-locality (i.e., Bell inequality violation) [6], to quantum steering [7] and, highly illuminating in my opinion, by the conceptual tool of the non-local PR-boxes that summarizes in a beautifully simple equation, $a \oplus b = x \cdot y$, the involved mathematical concept of entanglement [8]. Something analogous for joint measurements is still missing.

2. Quantum Teleportation and High-Impact Journals

On request of the editor, let me stress that "this section presents the author's own opinion regarding publication trends in quantum information" (see Endnote [9]).

Since the advent of quantum teleportation, especially since its first experimental demonstrations [10–12], it has become quasi-mandatory to publish in journals with high impact factors, like Nature, Nature Physics, Nature Photonics, Science and PRL. For example, all papers on long-distance quantum teleportation followed that trend (well, probably I am missing some, precisely those that do not follow that pattern): [13–18]. So, what happens if you resist the trend? We tried. We published an experiment in which the state to be teleported was carried by a photon produced long after the entangled photons constituting the quantum teleportation channel had left the laboratory. This required that the entangled photons and the photon carrying the state to be teleported were produced by different laser pulses (though from the same laser). This appeared in J. Opt. Soc. Am. B [19] and received a relatively low number of citations. This is the price to pay for independence. But who cares about independence today (see Endnote [20]—which refers to Reference [13])?

I am not complaining, but find it interesting to be aware of the huge impact quantum teleportation had on our community's trend to overvalue high-profile journals, with all the frustration that too often comes along. Unfortunately, that trend spread all over quantum information science. Admittedly, I am not the least responsible person for that (see Endnote [21]). Sorry.

3. The Bell-State Measurement in Quantum Networks

The joint measurement exploited in quantum teleportation, known as a Bell state measurement (BSM), is characterized by all its eigenvectors being maximally entangled. For instance, teleportation of qubits require the BSM whose eigenvectors are the four Bell states:

$$|\phi^{\pm}\rangle = (|0,0\rangle \pm |1,1\rangle)/\sqrt{2} \qquad (1)$$

$$|\psi^{\pm}\rangle = (|0,1\rangle \pm |1,0\rangle)/\sqrt{2} \qquad (2)$$

As already pointed out in the original paper [1], quantum teleportation can be extended to teleportation of entanglement, known as entanglement swapping. This, in turn, can be extended to teleportation over entire and complex networks [22], as illustrated in Figure 1. In such networks, each node with more than one edge performs a joint measurement, possibly on more than two systems. For simplicity, here we concentrate on only two cases, either a line or a triangle, see Figures 2 and 3. Notice that here only players with a single edge get inputs, denoted x and y, that determine which measurement to perform.

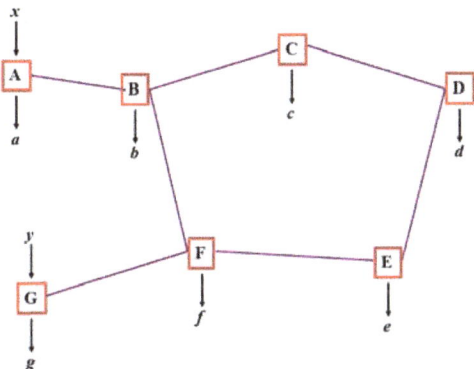

Figure 1. Example of a quantum network. Each edge represents a resource shared by the connected nodes. The resource are entangled quantum states, or, in order to compare with classical networks, correlated local variables (i.e., shared randomness). In this paper we consider only cases where inputs are provided to parties connected by a single edge.

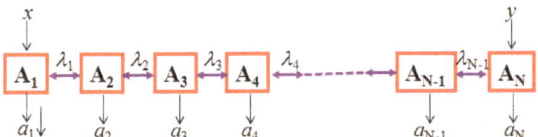

Figure 2. (N-1)-local scenario in a line [23]. The λ_j's represent independent quantum states, or, in the classical scenario used for comparison, random independent local variables. Only the first and last parties get inputs, x and y respectively.

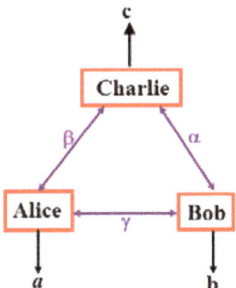

Figure 3. The triangle configuration for three parties [23]. Each pair of parties shares either a quantum state and performs quantum measurements—quantum scenario, or shares independent random variables α, β and γ and outputs a function of the random variables to which they have access. Notice that the three random variables are only used locally, hence the terminology three-local scenario. The "quantum grail" is to find a quantum scenario (without external inputs) leading to a probability $p(a,b,c)$ which can't be reproduced by any three-local scenario.

Let us first consider the triangle, see Figure 3. If Alice, Bob and Charlie each perform the BSM, then there is a simple classical model that reproduces the statistics of their outcomes, $p(a, b, c)$—notice that there are no inputs (see Endnote [24]—which refers to Reference [25]). Hence, somewhat surprisingly, in this case the joint measurement doesn't produce any quantum signature: such a triangle with BSM displays no quantumness.

Let's now consider the line of Figure 2. Start with only two edges. This corresponds to the scenario of entanglement swapping, i.e., of quantum teleportation of entanglement. For this simple case we name the parties with their usual names, i.e., Alice, Bob and Charlie, instead of A_1, A_2 and A_3, and similarly for the outcomes. Depending on Bob's outcome b, Alice's and Charlie's qubits get projected onto different entangled states; which exact entangled state depends on b. This can be checked with some entanglement witness, or, in a device-independent way, with some Bell inequality. For the Clauser-Horne-Shimony-Holt (CHSH) inequality, assuming perfect (noise-free) measurements, a violation is obtained if the product of the visibilities (see Endnote [26]) satisfies $W_1 \cdot W_2 > 1/\sqrt{2}$. In the symmetric case, $W_1 = W_2$, which implies $W_j > 2^{-1/4} \approx 84\%$. Such a high visibility has been achieved experimentally, e.g., [27], but with non-independent sources for the two quantum states ρ_1 and ρ_2 represented by the edges.

However, in such an entanglement scenario with independent sources, like e.g., [28], it is very natural to check for quantumness by comparing it with classical correlations under the assumption that the local (hidden) variables are also independent:

$$P(\lambda_1, \lambda_2) = P(\lambda_1) \cdot P(\lambda_2). \tag{3}$$

Such a case is called bi-local [23,29], to contrast it with the usual Bell locality. In case of n independent sources, the achievable classical correlations are called n-local [30–32].

In the bi-locality scenario it has been proven that a visibility product of $W_1 \cdot W_2 > \frac{1}{2}$ suffices to prove quantumness, i.e., to prove a quantum advantage over bi-local classical correlations [23,29]. Accordingly, in the symmetric case $W_j < 1/\sqrt{2} \approx 71\%$ suffices, as, e.g., in the experiment of Reference [28]. In this scenario, an explicit non-linear inequality (non-linear because the set of n-local correlations in non convex for all $n \geq 2$) has been found and fully analyzed [33]. The analyses show that this bi-local scenario is essentially identical to the old and well-known CHSH-Bell inequality between two parties. The relation builds on the fact that the two-bit outcome of the BSM is equivalent to the outcome of $\sigma_z \otimes \sigma_z$ for the first bit and $\sigma_x \otimes \sigma_x$ for the second bit. Hence, in a nutshell, Bob measures both of his qubits in the x–z bases, while Alice and Charlie measure in the $\pm 45°$ bases, exactly as in the CHSH case.

This is quite disappointing, as the threshold visibility per singlet, $1/\sqrt{2}$, is identical to the simpler case of CHSH between two parties. Apparently, the assumption of independent local variables λ_1 and λ_2 plays no role. But that cannot be! Independence is a strong assumption, it should thus lead to consequences. This illustrates how poorly we understand joint measurements. Could it be that increasing the number of inputs at Alice and Charlie's side, or studying longer linear chains, allows one to lower the threshold visibility per singlet? Reference [30], which considers n-locality in longer lines, and reference [31], which derives n-local inequalities from Bell inequalities, suggest the contrary and, so far, numerous numerical searches lead to disappointing results, see though the interesting findings in [34–37].

The mentioned negative results are no proof that the bi-local scenario is useless to lower the threshold visibility per singlet. But they call for alternative ideas. One nice idea is to go for a star network [31,38], though so far results seem very similar to the bi-local case.

The next section recalls results first presented in [39], a paper I never submitted to any journal, hence parts of it are reproduced here. In a nutshell, it presents another joint measurement and applies it to a three-partite scenario in the triangle configuration with three independent sources.

4. The Elegant Joint Measurement on Two Qubits

In order to study joint measurements different from the BSM we like to find a two-qubit basis with four partially entangled eigenstates, all with the same degree of entanglement and some nice symmetries. For this, we start with the four vertices of the tetrahedron inscribed in the Poincaré sphere:

$$\vec{m}_1 = (1,1,1)/\sqrt{3} \tag{4}$$
$$\vec{m}_2 = (1,-1,-1)/\sqrt{3} \tag{5}$$
$$\vec{m}_3 = (-1,1,-1)/\sqrt{3} \tag{6}$$
$$\vec{m}_4 = (-1,-1,1)/\sqrt{3} \tag{7}$$

Using cylindrical coordinates, $\vec{m}_j = (\sqrt{1-\eta_j^2}\cos\phi_j, \sqrt{1-\eta_j^2}\sin\phi_j, \eta_j)$, one obtains the natural correspondence with qubit states (note that here $\eta_j = \pm 1/\sqrt{3}$ for all j):

$$|\vec{m}_j\rangle = \sqrt{\frac{1-\eta_j}{2}} e^{i\phi_j/2}|0\rangle + \sqrt{\frac{1+\eta_j}{2}} e^{-i\phi_j/2}|1\rangle \tag{8}$$

Note that $\vec{m}_j = \langle \vec{m}_j | \vec{\sigma} | \vec{m}_j \rangle$, as expected (with $\vec{\sigma}$ the three Pauli matrices).

Inspired by [40,41], we consider the following 2-qubit basis constructed on anti-parallel spins [39]:

$$|\Phi_j\rangle = \sqrt{\frac{3}{2}}|\vec{m}_j, -\vec{m}_j\rangle + i\frac{\sqrt{3}-1}{2}|\psi^-\rangle \tag{9}$$
$$= \frac{\sqrt{3}+1}{2\sqrt{2}}|\vec{m}_j, -\vec{m}_j\rangle + \frac{\sqrt{3}-1}{2\sqrt{2}}|-\vec{m}_j, \vec{m}_j\rangle, \tag{10}$$

where $|-\vec{m}\rangle$ is orthogonal to $|\vec{m}\rangle$: it has the same form as (8) but with $\eta \to -\eta$ and $\phi \to \phi + \pi$. Notice that in (10) the states Φ_j are written in their Schmidt bases.

In order to check that the Φ_j are normalized and mutually orthogonal one should use $\langle \vec{m}, -\vec{m} | \psi^- \rangle = i/\sqrt{2}$ for all $\vec{m}$ and $\langle \vec{m}_j, -\vec{m}_j | \vec{m}_k, -\vec{m}_k \rangle = 1/3$ for all $j \neq k$.

Using the corresponding one-dimensional projectors:

$$|\Phi_j\rangle\langle\Phi_j| = \frac{1}{4}\left(\mathbb{1} + \frac{\sqrt{3}}{2}(\vec{m}_j\vec{\sigma}\otimes\mathbb{1} - \mathbb{1}\otimes\vec{m}_j\vec{\sigma}) - \frac{3}{2}\sum_{n,k}m_{j,n}m_{j,k}\sigma_n\otimes\sigma_k + \frac{1}{2}\vec{\sigma}\otimes\vec{\sigma}\right), \tag{11}$$

it is not difficult to compute the partial traces and observe the elegant properties:

$$\langle\Phi_j|\vec{\sigma}\otimes\mathbb{1}|\Phi_j\rangle = \frac{1}{2}\vec{m}_j \tag{12}$$
$$\langle\Phi_j|\mathbb{1}\otimes\vec{\sigma}|\Phi_j\rangle = -\frac{1}{2}\vec{m}_j. \tag{13}$$

In words, the partial states (obtained by tracing out one party) point along the edges of the tetrahedron, but with Bloch vectors of reduced lengths $\frac{1}{2}$.

We name the two-qubit measurement with eigenstates (9) and (10) the elegant joint measurement (EJM). We believe it is unique with all four eigenstates having identical degrees of partial entanglement and with all partial states of all eigenstates parallel or anti-parallel to the vertices of the tetrahedron.

5. Quantum Correlation from Singlets and the EJM in the Triangle Configuration

Consider three independent singlets in the triangle configuration and assume that Alice, Bob and Charlie each perform the EJM on their two (independent) qubits, see Figure 3. Denote the resulting correlation $p_{tr}(a,b,c)$, where $a,b,c = 1,2,3,4$. By symmetry, $p_{tr}(a,b,c)$ is fully characterized by three numbers corresponding to the cases $a = b = c$, $a = b \neq c$ (and circular permutations, i.e., two outcomes are equal, but the third differs) and $a \neq b \neq c \neq a$. A not too complex computation gives [39]:

$$p_{tr}(a = k, b = k, c = k) = \frac{25}{256} \text{ for } k = 1, 2, 3, 4 \tag{14}$$

$$p_{tr}(a = k, b = k, c = m) = \frac{1}{256} \text{ for } k \neq m \tag{15}$$

$$p_{tr}(a = k, b = n, c = m) = \frac{5}{256} \text{ for } k \neq n \neq m \neq k. \tag{16}$$

The normalization holds: $4 \cdot \frac{25}{256} + 36 \cdot \frac{1}{256} + 24 \cdot \frac{5}{256} = 1$.

As expected $p_{tr}(a) = p_{tr}(b) = p_{tr}(c) = \frac{1}{4}$. More interesting is the probabilities that two parties get identical results:

$$\begin{aligned} p_{tr}(a = k, b = k) &= p_{tr}(a = b = c = k) + p_{tr}(a = b = k, c \neq k) \\ &= \frac{25 + 3 \cdot 1}{256} = \frac{7}{64}. \end{aligned} \tag{17}$$

Hence, all pairs of parties are correlated, e.g., $p_{tr}(a|b) \neq \frac{1}{4}$. In worlds, given an outcome $b = k$ for Bob, Alice's outcome has a large chance to take the same value: $p_{tr}(a = k|b = k) = \frac{p_{tr}(a=k,b=k)}{p_{tr}(b=k)} = \frac{7}{16}$. Accordingly:

$$p_{tr}(a = b) = \sum_k p_{tr}(b = k) p(a = k|b = k) = \frac{7}{16}. \tag{18}$$

The strength of the three-party correlation is even more impressive:

$$p_{tr}(a = k|b = c = k) = \frac{p_{tr}(a = b = c = k)}{p_{tr}(b = c = k)} = \frac{25}{28}. \tag{19}$$

Hence $p_{tr}(a = b = c) = 4 \cdot \frac{25}{256} = \frac{25}{64}$.

The high correlation displayed by p_{tr} strongly suggests that it can't be realized by any three-local model. However, one has to be careful. Indeed, reference [39] presents two three-local models with even higher correlations, though not symmetric and not reproducing the correlations (14)–(16) of p_{tr}. For completeness, these two models are reproduced in the next Section 6. Since [39] was posted on the arXiv quite some researchers tried to prove or disprove the three-local nature of p_{tr}. In particular Elisa Bäumer and Elie Wolfe (private communications) devoted time to this fascinating question, the first one with strong arguments in favour of a negative answer and the second one, using his "inflation method" [42,43], arguing in favour of a positive answer. The fact is that the three-local nature of p_{tr} remains elusive. More generally, the existence/nonexistence of a quantum scenario that can provably not be reproduced by any three-local model and that respects the triangle symmetry, or some other closed symmetric loop, remains open, illustrating how poorly we understand joint measurements. Let me emphasize that if such a quantum example exists, its quantumness could only be due to the joint measurements, as in a loop there are no "ends", hence no parties with inputs, in strong contrast to the by now common Bell inequality scenarios. I elaborate on this in Section 7.

6. Is $p_{tr}(a, b, c)$ Three-Local?

In this section, we consider the question whether the quantum probability $p_{tr}(a, b, c)$ is three-local, i.e., whether it can be reproduced by a 3-local model:

$$p_{tr} \stackrel{?}{=} \sum_{\alpha\beta\gamma} P(\alpha) P(\beta) P(\gamma) P(a|\beta, \gamma) P(b|\gamma, \alpha) P(c|\alpha, \beta). \tag{20}$$

In such a three-local model of $p_{tr}(a,b,c)$ the Alice–Bob correlation could only be due to their shared randomness γ. Similarly, the correlation between Bob and Charlie is necessarily due to α and the Alice–Charlie correlation due to β. Accordingly, each local variable α, β and γ would contain a four-dit, equally distributed among the values 1, 2, 3, 4, and with a relatively high probability both Alice and Bob output the four-digit contained in γ, and similarly for the other pairs of parties. Admittedly, this is only an argument, not a proof of the conjecture that p_{tr} is non-local.

Accordingly, let's consider the following natural type of three-local models. Let $\gamma = (\gamma_1, \gamma_2)$, where $\gamma_1 = 1, 2, 3, 4$, each with equal probability and $\gamma_2 = 0, 1$ with $prob(\gamma_2 = 1) = q$. The idea is that whenever $\gamma_2 = 1$, then Alice and Bob results are given by γ_1, hence Alice and Bob get perfectly correlated. More explicitly, Alice's output function reads:

$$a(\beta, \gamma) = \begin{cases} \gamma_1 & \text{if } \beta_2 = 0 \text{ and } \gamma_2 = 1 \\ \beta_1 & \text{if } \beta_2 = 1 \text{ and } \gamma_2 = 0 \\ \beta_1|\gamma_1 & \text{if } \beta_2 = \gamma_2 \end{cases}, \qquad (21)$$

where $\beta_1|\gamma_1$ indicates that $a(\beta, \gamma)$ equals β_1 or γ_1 with equal probability $\frac{1}{2}$.

Table 1 indicates all possible outputs (where $\bar{q} \equiv (1-q) = prob(\alpha_2 = 0) = prob(\beta_2 = 0) = prob(\gamma_2 = 0)$).

Averaging the probabilities that $a = b = c$ over the eight combinations of values of α_2, β_2 and γ_2, i.e., over the eight lines of Table 1, gives:

$$\begin{aligned} p_{3loc}(a = b = c) &= \frac{13}{64}(\bar{q}^3 + q^3) + \frac{3}{4}(\bar{q}^2 q + \bar{q} q^2) \\ &= \frac{13 + 9q - 9q^2}{64} \end{aligned} \qquad (22)$$

Table 1. The eight lines correspond to the eight possible combinations of values of α_2, β_2 and γ_2 (first three columns). The next three columns indicate Alice, Bob and Charlie's outputs. The seventh column indicates the probability of the corresponding line and the last two columns the probability that $a = b$ and $a = b = c$, respectively.

α_2	β_2	γ_2	a	b	c	P	Prob $(a = b)$	Prob $(a = b = c)$
0	0	0	$\beta_1|\gamma_1$	$\alpha_1|\gamma_1$	$\alpha_1|\beta_1$	$\bar{q}^3$	7/16	13/64
0	0	1	γ_1	γ_1	$\alpha_1|\beta_1$	$\bar{q}^2 q$	1	1/4
0	1	0	β_1	$\alpha_1|\gamma_1$	β_1	$\bar{q}^2 q$	1/4	1/4
0	1	1	$\beta_1|\gamma_1$	γ_1	β_1	$\bar{q} q^2$	5/8	1/4
1	0	0	$\beta_1|\gamma_1$	α_1	α_1	$\bar{q}^2 q$	1/4	1/4
1	0	1	γ_1	$\alpha_1|\gamma_1$	α_1	$\bar{q} q^2$	5/8	1/4
1	1	0	β_1	α_1	$\alpha_1|\beta_1$	$\bar{q} q^2$	1/4	1/4
1	1	1	$\beta_1|\gamma_1$	$\alpha_1|\gamma_1$	$\alpha_1|\beta_1$	q^3	7/16	13/64

Hence, the maximal three-particle correlation of our three-local model is achieved for $q = \frac{1}{2}$ and reads:

$$\max_q p_{3loc}(a = b = c) = \frac{61}{256} \qquad (23)$$

This is much smaller than the value obtained in the quantum case with the elegant joint measurement.

The above is not a proof, but leads us to conjecture that the quantum probability $p_{tr}(a, b, c)$ is not three-local. Indeed, γ has to correlate A and B, i.e., γ contributes to the probability that $a = b$, and β contributes to $p_{tr}(a = c)$ and α contributes to $p_{tr}(b = c)$. But then the three independent variables α, β and γ can't do the job for the three-particle correlation $a = b = c$.

Note that if the outcomes are grouped two by two, such that outcomes are binary, then a three-local model similar to (21) can reproduce the quantum correlation. But, again, with four outcomes per party this seems impossible.

A Natural but Asymmetric Three-Local Model

There is another three-local model that we need to consider, directly inspired by the quantum singlet states shared by each pair of parties. Assume that the three local variables α, β and γ each take values (0, 1) or (1, 0) with 50% probabilities, where the first bit of α is sent to Bob and the second bit to Charlie, and similarly for β and γ. Clearly, this three-local model assumes binary local variables, i.e., bits, but we like to keep the notation (0, 1) and (1, 0) for the two values.

The outcomes are then determined by the two bits that each party receives from the local variables it shares with his two neighbours. We like to maximize the probability $p(a = b = c)$. All output functions that maximize $p(a = b = c)$ are equivalent. One possible choice is:

$$(0,0) \Rightarrow a = 2, \ b = 4, \ c = 3 \tag{24}$$
$$(0,1) \Rightarrow a = 1, \ b = 1, \ c = 1 \tag{25}$$
$$(1,0) \Rightarrow a = 3, \ b = 2, \ c = 4 \tag{26}$$
$$(1,1) \Rightarrow a = 4, \ b = 3, \ c = 2 \tag{27}$$

Note that in this three-local model γ imposes that both Alice and Bob can only output one out of two values. Which of the two values happens depends on the second local variable. This provides intuition as to why this three-local model achieves $p(a = b = c) = \frac{1}{2}$, i.e., an even larger value than the quantum probabilities with the EJM. Moreover $p(a = b) = \frac{1}{2}$, hence $p(a = b = c|a = b) = 1$. However, this model does not respect the symmetries of the quantum scenario. In particular 20 out of the 24 cases $p(a = k, b = n, c = m)$ with $k \neq n \neq m \neq k$ take values 0 (recall that in the quantum scenario all 24 probabilities take value $\frac{5}{256}$, see Equations (14)–(16)).

This simple three-local model shows that in order to prove the non-three-locality of $p_{tr}(a,b,c)$ it is not sufficient to consider $p(a = b = c)$, but one has to consider also the cases $a \neq b \neq c$.

7. Consequences of a Non-Three-Local Quantum Triangle

Let's assume that there is a nicely symmetric quantum example of a triangle provably not three-local, e.g., a probability distribution $p(a,b,c)$ which derives from three independent quantum states and identical quantum measurements in the triangle configuration, see Figure 3, that has no three-local decomposition (20) (see Endnote [44]—which refers to References [24,25,45]). What would that imply for our worldview? First, notice that in such a scenario there are no inputs. Accordingly, one could imagine a toy universe consisting of only six qubits, without anything outside, which nevertheless manifests quantumness, including provable randomness. Well, the outcomes a, b and c should get out of this mini-quantum-universe in order to produce any evidence; one more manifestation of the infamous quantum measurement problem [46,47]. This is in strong contrast to the usual Bell inequality scenario where inputs provided from outside the systems under test are essential to prove any quantumness. Of course, our six qubit toy universe must satisfy the assumption of independence of the three sources (without any assumption, nothing can be proven). But this assumption is really minimal: if the sources are spatially separated, then it is very natural to assume that they are independent. The first source could be powered by solar power and produce entangled photons, the second source powered by human energy and produce entangled atoms, and the third source powered by nuclear power and produce some entangled quantum "stuff", e.g., cats or crystals [48].

Admittedly, one may argue that Alice, for instance, somehow gets inputs from the sources denoted β and γ on Figure 3. But in Bell inequality scenarios, one never thinks of the source in-between Alice and Bob as the inputs, the inputs are determining the measurement setting and, in Bell scenarios, necessarily come from outside the quantum systems. Nothing like this in the triangle scenario. Quantumness would be proven from inside the six qubit toy universe (see Endnote [49]—which refers to Reference [23]). Also quantum randomness would be proven within this toy universe.

A second interesting consequence of a "quantum triangle" appears when one moves the sources α, β and γ close to one of the players, or even inside the players. Assume the source α is given to Bob, β is given to Charlie and γ to Alice. In the quantum case, Alice, Bob and Charlie each emits some quantum state, e.g., one qubit, and sends it to his partner counter-clock wise. In the classical case they each send an arbitrarily large amount of classical information (possibly infinite) to their partner, still counter-clock wise. The three-local assumption of independence translates into the assumption that all communications are well enough synchronized to guarantee that each party sends out his quantum state or classical information before receiving anything from his partner. In this way one compares the power of quantum communication (of even just a qubit) with the power of classical communication, possibly an infinite amount of classical information. Under the synchronization assumption of the communications, one would prove the superiority of the former over the latter.

Admittedly, a similar story of replacing entanglement (shared randomness) by quantum (classical) communication can be told for the standard Bell inequality scenario. Instead of an entanglement source in-between Alice and Bob, Alice would send a quantum state to Bob prior to receiving her input x. This would allow them to violate the CHSH-Bell inequality, while if Alice is restricted to sending classical information—prior to receiving her input—they can't violate any Bell inequality.

8. Conclusions

In summary, 25 years after the beautiful invention of quantum teleportation lots of progress has been made on Bell-locality [6], on quantum steering [7] and more generally quantum information theory. Likewise enormous progress happens in experimental, applied and engineering, even in industrialization of quantum technologies [50–52]. But, quite surprisingly and disappointingly, essentially no progress took place in improving our understanding of joint measurements (see Endnote [53]—which refers to Reference [54–57]), i.e., on the second usage of entanglement in quantum teleportation. For example, it was proven that there is no simple analog of PR-boxes for joint measurements [58–61]. This is exciting, as it indicates that big surprises still await us in the—hopefully not too far—future.

Funding: This work partially supported by the Swiss NCCR-QSIT and the European ERC-AG MEC.

Acknowledgments: The present version profited from valuable comments by Nicolas Brunner, Sandu Popescu, Armin Tavakoli and—for once—the two referees.

Conflicts of Interest: The author declares no conflict of interest.

References

1. Bennett, C.H.; Brassard, G.; Crépeau, C.; Jozsa, R.; Peres, A.; Wootters, W.K. Teleporting an unknown quantum state via dual classical and Einstein-Podolsky-Rosen channels. *Phys. Rev. Lett.* **1993**, *70*, 1895. [CrossRef] [PubMed]
2. Pirandola, S.; Eisert, J.; Weedbrook, C.; Furusawa, A.; Braunstein, S.L. Advances in quantum teleportation. *Nat. Photonics* **2015**, *9*, 641. [CrossRef]
3. Horodecki, R.; Horodecki, P.; Horodecki, M.; Horodecki, K. Quantum entanglement. *Rev. Mod. Phys.* **2009**, *81*, 865. [CrossRef]
4. Gisin, N. *Quantum Chance, Nonlocality, Teleportation and Other Quantum Marvels*; Springer: Berlin, Germany, 2014.
5. As said, this is only an intuitive explanation, as there are no two-qubit states with this property. For a more formal description of quantum teleportation see [1,2], though the here presented intuition contains the essential point in the present context.
6. Brunner, N.; Cavalcanti, D.; Pironio, S.; Scarani, V.; Wehner, S. Bell nonlocality. *Rev. Mod. Phys.* **2014**, *86*, 419. [CrossRef]

7. Cavalcanti, D.; Skrzypczyk, P. Quantum steering: A review with focus on semidefinite programming. *Rep. Prog. Phys.* **2017**, *80*, 024001. [CrossRef]
8. Popescu, S.; Rohrlich, D. Quantum nonlocality as an axiom. *Found. Phys.* **1994**, *24*, 379–385. [CrossRef]
9. Let me add that this is true of all opinions expressed in all my papers.
10. Boschi, D.; Branca, S.; de Martini, F.; Hardy, L.; Popescu, S. Experimental realisation of teleporting an unknown pure quantum state via dual classical and Einstein-Podolski-Rosen channels. *Phys. Rev. Lett.* **1998**, *80*, 1121. [CrossRef]
11. Bouwmeester, D.; Pan, J.-W.; Mattle, K.; Eibl, M.; Weinfurter, H.; Zeilinger, A. Experimental quantum teleportation. *Nature* **1997**, *390*, 575. [CrossRef]
12. Furusawa, A.; Sørensen, J.L.; Braunstein, S.L.; Fuchs, C.A.; Kimble, H.J.; Polzik, E.S. Unconditional quantum teleportation. *Science* **1998**, *282*, 706–709. [CrossRef]
13. Marcikic, I.; de Riedmatten, H.; Tittel, W.; Zbinden, H.; Gisin, N. Long-distance teleportation of qubits at telecommunication wavelengths. *Nature* **2003**, *421*, 509. [CrossRef]
14. Ursin, R.; Jennewein, T.; Aspelmeyer, M.; Kaltenbaek, R.; Lindenthal, M.; Walther, P.; Zeilinger, A. Quantum teleportation across the Danube. *Nature* **2004**, *430*, 849. [CrossRef]
15. Yin, J.; Ren, J.G.; Lu, H.; Cao, Y.; Yong, H.L.; Wu, Y.P.; Liu, C.; Liao, S.K.; Zhou, F.; Jiang, Y.; et al. Quantum teleportation and entanglement distribution over 100 km free space channels. *Nature* **2012**, *488*, 185–188. [CrossRef]
16. Ma, X.S.; Herbst, T.; Scheidl, T.; Wang, D.; Kropatschek, S.; Naylor, W.; Wittmann, B.; Mech, A.; Kofler, J.; Anisimova, E.; et al. Quantum teleportation over 143 km using active feed forward. *Nature* **2012**, *489*, 269–273. [CrossRef] [PubMed]
17. Ren, J.G.; Xu, P.; Yong, H.L.; Zhang, L.; Liao, S.K.; Yin, J.; Liu, W.Y.; Cai, W.Q.; Yang, M.; Li, L.; et al. Ground to satelite quantum teleportation. *Nature* **2017**, *549*, 70–73. [CrossRef] [PubMed]
18. Bussières, F.; Clausen, C.; Tiranov, A.; Korzh, B.; Verma, V.B.; Nam, S.W.; Marsili, F.; Ferrier, A.; Goldner, P.; Herrmann, H.; et al. Quantum teleportation from a telecom-wavelength photon to a solid-state quantum memory. *Nat. Photonics* **2014**, *8*, 775–778. [CrossRef]
19. Landry, O.; van Houwelingen, J.A.W.; Beveratos, A.; Zbinden, H.; Gisin, N. Quantum teleportation over the Swisscom telecommunication network. *J. Opt. Soc. Am. B* **2007**, *24*, 398–403. [CrossRef]
20. Here is an instructive example. I wrote (among others) the introduction to our long-distance quantum teleportation paper [13] and cited Aristotle for his distinction of form and substance that make up objects. When the proofs arrived we discovered that the editor dared to remove all this stuff about Aristotle, form and substance (although she/he is not a co-author of our paper, isn't it?) I got angry and suggested to my students to withdraw our (accepted) submission to Nature. That proposal triggered a sort of nuclear bomb. No way to argue against the dominant fashion. I surrendered. But the arXiv version of our paper still contains Aristotle (quant-ph/0301178).
21. Though, before having students I used to send all my papers to Physics Letters A, a journal with the enormous quality of always accepting all my submissions, hence allowing me to concentrate on research.
22. Briegle, H.-J.; Dur, W.; Cirac, J.I.; Zoller, P. Quantum repeaters: The role of imperfect local operations in quantum communication. *Phys. Rev. Lett.* **1998**, *81*, 5932–5935. [CrossRef]
23. Branciard, C.; Rosset, D.; Gisin, N.; Pironio, S. Bilocal versus nonbilocal correlations in entanglement-swapping experiments. *Phys. Rev. A* **2012**, *85*, 032119. [CrossRef]
24. Another three-partite scenario in a triangle configuration without inputs should be mentioned here [25], though it is essentially the usual Clauser-Horne-Shimony-Holt (CHSH) two-party case with the two random number generators collected as the third party. As expected, the resistance to noise per singlet is poor, certainly not better than for the usual CHSH inequality.
25. Fritz, T. Beyond Bell's theorem: Correlation scenarios. *New J. Phys.* **2012**, *14*, 103001. [CrossRef]
26. Recall that for the Werner states $\rho_W = W \cdot |\psi^-\rangle\langle\psi^-| + (1-W)\mathbb{1}/4$, where $|\psi^-\rangle$ denotes the singlet, the visibility equals W.
27. Jennewein, T.; Weihs, G.; Pan, J.-W.; Zeilinger, A. Experimental nonlocality proof of quantum teleportation and entanglement swapping. *Phys. Rev. Lett.* **2002**, *88*, 017903. [CrossRef]
28. Halder, M.; Beveratos, A.; Gisin, N.; Scarani, V.; Simon, C.; Zbinden, H. Entangling independent photons by time measurement. *Nat. Phys.* **2007**, *3*, 692–695. [CrossRef]

29. Branciard, C.; Gisin, N.; Pironio, S. Characterizing the Nonlocal Correlations Created via Entanglement Swapping. *Phys. Rev. Lett.* **2010**, *104*, 170401.
30. Rosset, D.; Branciard, C.; Barnea, T.J.; Pütz, G.; Brunner, N.; Gisin, N. Nonlinear Bell Inequalities Tailored for Quantum Networks. *Phys. Rev. Lett.* **2016**, *116*, 010403. [CrossRef]
31. Tavakoli, A.; Renou, M.-O.; Gisin, N.; Brunner, N. Correlations in star networks: From Bell inequalities to network inequalities. *New J. Phys.* **2017**, *19*, 073003. [CrossRef]
32. Tavakoli, A. Quantum correlations in connected multipartite Bell experiments. *J. Phys. A* **2016**, *49*, 145304. [CrossRef]
33. Gisin, N.; Mei, Q.; Tavakoli, A.; Renou, M.-O.; Brunner, N. All entangled pure quantum states violate the bilocality inequality. *Phys. Rev. A* **2017**, *96*, 020304. [CrossRef]
34. Chaves, R. Polynomial Bell Inequalities. *Phys. Rev. Lett.* **2016**, *116*, 010402. [CrossRef] [PubMed]
35. Fraser, T.C.; Wolfe, E. Causal compatibility inequalities admitting quantum violations in the triangle structure. *Phys. Rev. A* **2018**, *98*, 022113. [CrossRef]
36. Fritz, T. Beyond Bell's Theorem II: Scenarios with Arbitrary Causal Structure. *Commun. Math. Phys.* **2016**, *341*, 391–434. [CrossRef]
37. Andreoli, F.; Carvacho, G.; Santodonato, L.; Chaves, R.; Sciarrino, F. Maximal qubit violation of n-locality inequalities in a star-shaped quantum network. *New J. Phys.* **2017**, *19*, 113020. [CrossRef]
38. Tavakoli, A.; Skrzypczyk, P.; Cavalcanti, D.; Acín, A. Nonlocal correlations in the star-network configuration. *Phys. Rev. A* **2014**, *90*, 062109. [CrossRef]
39. Gisin, N. The elegant joint quantum measurement and some conjectures about N-locality in the triangle and other configurations. *arXiv* **2017**, arXiv:1708.05556.
40. Massar, S.; Popescu, S. Optimal Extraction of Information from Finite Quantum Ensembles. *Phys. Rev. Lett.* **1995**, *74*, 1259. [CrossRef]
41. Gisin, N.; Popescu, S. Spin Flips and Quantum Information for Antiparallel Spins. *Phys. Rev. Lett.* **1999**, *83*, 432. [CrossRef]
42. Wolfe, E.; Spekkens, R.W.; Fritz, T. The Inflation Technique for Causal Inference with Latent Variables. *arXiv* **2018**, arXiv:1609.00672.
43. Navascues, M.; Wolfe, E. The inflation technique solves completely the classical inference problem. *arXiv* **2017**, arXiv:1707.06476.
44. While finishing this work, Prof. Salman Beigi sent me what appears to be the first such example [45]! For a non-symmetric example see Fritz's example [25] recalled in Note [24].
45. Beigi, S. (Institute for Research in Fundamental Science (IPM), Tehran, Iran). Personal communication, 2018.
46. Gisin, N. Collapse. What else? In *Collapse of the Wave Function*; Gao, S., Ed.; Cambridge University Press: Cambridge, UK, 2018.
47. Gisin, N.; Fröwis, F. From quantum foundations to applications and back. *Philos. Trans. R. Soc. A* **2018**, *376*. [CrossRef]
48. Usmani, I.; Clausen, C.; Bussières, F.; Sangouard, N.; Afzelius, M.; Gisin, N. Heralded quantum entanglement between two crystals. *Nat. Photonics* **2012**, *6*, 234. [CrossRef]
49. Note that the six qubits could also be on a line, as in Figure 7 of [23].
50. Available online: www.idquantique.com (accessed on 27 September 2018).
51. Available online: www.idquantique.com/introducing-quantum-rng-chip (accessed on 27 September 2018).
52. Boaron, A.; Boso, G.; Rusca, D.; Vulliez, C.; Autebert, C.; Caloz, M.; Perrenoud, M.; Gras, G.; Bussières, F.; Li, M.J.; et al. Secure quantum key distribution over 421 km of optical fiber. *Phys. Rev. Lett.* **2018**, *121*, 190502. [CrossRef] [PubMed]
53. One exception is the possibility of detecting joint measurements in a device-independent way, see, e.g., [54–57].
54. Rabelo, R.; Ho, M.; Cavalcanti, D.; Brunner, N.; Scarani, V. Device-Independent Certification of Entangled Measurements. *Phys. Rev. Lett.* **2011**, *107*, 050502. [CrossRef] [PubMed]
55. Ciarán, M.L. Device-independent certification of non-classical measurements via causal models. *arXiv* **2018**, arXiv:1806.10895.
56. Renou, M.-O.; Kaniewski, J.; Brunner, N. Self-testing entangled measurements in quantum networks. *Phys. Rev. Lett.* **2018**, *121*, 250507. [CrossRef] [PubMed]

57. Bancal, J.-D.; Sangouard, N.; Sekatski, P. Noise-resistant device-independent certification of Bell state measurements. *Phys. Rev. Lett.* **2018**, *121*, 250506. [CrossRef]
58. Short, A.J.; Popescu, S.; Gisin, N. Entanglement swapping for generalized nonlocal correlations. *Phys. Rev. A* **2006**, *73*, 012101. [CrossRef]
59. Barrett, J. Information processing in generalized probabilistic theories. *Phys. Rev. A* **2007**, *75*, 032304. [CrossRef]
60. Skrzypczyk, P.; Brunner, N. Couplers for non-locality swapping. *New J. Phys.* **2009**, *11*, 073014. [CrossRef]
61. Skrzypczyk, P.; Brunner, N.; Popescu, S. Emergence of quantum correlations from nonlocality swapping. *Phys. Rev. Lett.* **2009**, *102*, 110402. [CrossRef] [PubMed]

© 2019 by the authors. Licensee MDPI, Basel, Switzerland. This article is an open access article distributed under the terms and conditions of the Creative Commons Attribution (CC BY) license (http://creativecommons.org/licenses/by/4.0/).

Article

Remote Sampling with Applications to General Entanglement Simulation

Gilles Brassard [1,2,*], Luc Devroye [3] and Claude Gravel [4,†,*]

1. Département d'Informatique et de Recherche Opérationnelle, Université de Montréal, Montréal, QC H3C 3J7, Canada
2. Canadian Institute for Advanced Research, Toronto, ON M5G 1M1, Canada
3. School of Computer Science, McGill University, Montréal, QC H3A 0E9, Canada; lucdevroye@gmail.com
4. National Institute of Informatics, 2-1-2 Hitotsubashi, Chiyoda, Tokyo 101-0003, Japan
* Correspondence: brassard@iro.umontreal.ca (G.B.); claudegravel1980@gmail.com (C.G.)
† Work performed at Université de Montréal.

Received: 13 June 2018; Accepted: 15 January 2019; Published: 19 January 2019

Abstract: We show how to sample exactly discrete probability distributions whose defining parameters are distributed among remote parties. For this purpose, von Neumann's rejection algorithm is turned into a distributed sampling communication protocol. We study the expected number of bits communicated among the parties and also exhibit a trade-off between the number of rounds of the rejection algorithm and the number of bits transmitted in the initial phase. Finally, we apply remote sampling to the simulation of quantum entanglement in its essentially most general form possible, when an arbitrary finite number m of parties share systems of arbitrary finite dimensions on which they apply arbitrary measurements (not restricted to being projective measurements, but restricted to finitely many possible outcomes). In case the dimension of the systems and the number of possible outcomes per party are bounded by a constant, it suffices to communicate an expected $O(m^2)$ bits in order to simulate *exactly* the outcomes that these measurements would have produced on those systems.

Keywords: communication complexity; quantum theory; classical simulation of entanglement; exact sampling; random bit model; entropy

---

1. Introduction

Let $\mathbb{X}$ be a nonempty finite set containing n elements and $p = (p_x)_{x \in \mathbb{X}}$ be a probability vector parameterized by some vector $\theta = (\theta_1, \ldots, \theta_m) \in \Theta^m$ for an integer $m \geq 2$. For instance, the set Θ can be the real interval $[0,1]$ or the set of Hermitian semi-definite positive matrices as it is the case for the simulation of entanglement. The probability vector p defines a random variable X such that $\mathbf{P}\{X = x\} \stackrel{\text{def}}{=} p_x$ for $x \in \mathbb{X}$. To sample exactly the probability vector p means to produce an output x such that $\mathbf{P}\{X = x\} = p_x$. The problem of sampling probability distributions has been studied and is still studied extensively within different random and computational models. Here, we are interested in sampling *exactly* a discrete distribution whose defining parameters are distributed among m different parties. The θ_i's for $i \in \{1, \ldots, m\}$ are stored in m different locations where the ith party holds θ_i. In general, any communication topology between the parties would be allowed, but, in this work, we concentrate for simplicity on a model in which we add a designated party known as the *leader*, whereas the m other parties are known as the *custodians* because each of them is sole keeper of the corresponding parameter θ—hence there are $m+1$ parties in total. The leader communicates in both directions with the custodians, who do not communicate among themselves. Allowing inter-custodian communication would not improve the communication efficiency of our scheme and can, at best, halve the number of bits communicated in any protocol. However, it could dramatically improve the sampling *time* in a realistic model in which each party is limited to sending and receiving a fixed number of bits at any given time step, as demonstrated in our previous work [1] concerning a special case of the problem considered here. The communication scheme is illustrated in Figure 1.

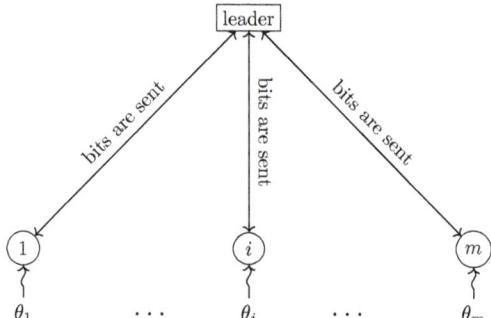

Figure 1. The communication scheme.

It may seem paradoxical that the leader can sample *exactly* the probability vector p with a *finite* expected number of bits sent by the custodians, who may hold *continuous* parameters that define p. However, this counterintuitive possibility has been known to be achievable for more than a quarter-century in earlier work on the simulation of quantum entanglement by classical communication, starting with Refs. [2–7], continuing with Refs. [8–14], etc. for the bipartite case and Refs. [15–17], etc. for the multipartite case, and culminating with our own Ref. [1].

Our protocol to sample remotely a given probability vector is presented in Section 2. For this purpose, the von Neumann rejection algorithm [18] is modified to produce an output $x \in \mathbb{X}$ with exact probability p_x using mere approximations of those probabilities, which are computed based on partial knowledge of the parameters transmitted on demand by the custodians to the leader. For the sake of simplicity, and to concentrate on the new techniques, we assume initially that algebraic operations on real numbers can be carried out with infinite precision and that continuous random variables can be sampled. Later, in Section 4, we build on techniques developed in Ref. [1] to obtain exact sampling in a realistic scenario in which all computations are performed with finite precision and the only source of randomness comes from flipping independent fair coins.

In the intervening Section 3, we study our motivating application of remote sampling, which is the simulation of quantum entanglement using classical resources and classical communication. Readers who may not be interested in quantum information can still benefit from Section 2 and most of Section 4, which make no reference to quantum theory in order to explain our general remote sampling strategies. A special case of remote sampling has been used by the authors [1], in which the aim was to sample a specific probability distribution appearing often in quantum information science, namely the m-partite Greenberger–Horne–Zeilinger (GHZ) distribution [19]. More generally, consider a quantum system of dimension $d = d_1 \cdots d_m$ represented by a density matrix ρ known by the leader (surprisingly, the custodians have no need to know ρ). Suppose that there are m generalized measurements (POVMs) acting on quantum systems of dimensions $d_1, \ldots, d_m$ whose possible outcomes lie in sets $\mathbb{X}_1, \ldots, \mathbb{X}_m$ of cardinality $n_1, \ldots, n_m$, respectively. Each custodian knows one and only one of the POVMs and nothing else about the others. The leader does not know initially any information about any of the POVMs. Suppose in addition that the leader can generate independent identically distributed uniform random variables on the real interval $[0, 1]$. We show how to generate a random vector $X = (X_1, \ldots, X_m) \in \mathbb{X} = \mathbb{X}_1 \times \ldots \times \mathbb{X}_m$ sampled from the exact joint probability distribution that would be obtained if each custodian i had the ith share of ρ (of dimension d_i) and measured it according to the ith POVM, producing outcome $x_i \in \mathbb{X}_i$. This task is defined formally in Section 3, where we prove that the total expected number of bits transmitted between the leader and the custodians using remote sampling is $O(m^2)$ provided all the d_i's and n_i's are bounded by some constant. The exact formula, involving m as well as the d_i's and n_i's, is given as Equation (14) in Section 3, where d and n denote the product of the d_i's and the n_i's, respectively. In Section 4, we obtain the same asymptotic result in the more realistic scenario in which the only source of randomness comes from independent identically distributed uniform random *bits*. This result subsumes that of Ref. [1] since all d_i's and n_i's are equal to 2 for projective measurements on individual qubits of the m-partite GHZ state.

2. Remote Sampling

As explained in the Introduction, we show how to sample *remotely* a discrete probability vector $p = (p_x)_{x \in \mathbb{X}}$. The task of sampling is carried by a *leader* ignorant of some parameters $\theta = (\theta_1, \ldots, \theta_m)$ that come in the definition of the probability vector, where each θ_i is known by the ith *custodian* only, with whom the leader can communicate. We strive to minimize the amount of communication required to achieve this task.

To solve our conundrum, we modify the von Neumann rejection algorithm [18,20]. Before explaining those modifications, let us review the original algorithm. Let $q = (q_x)_{x \in \mathbb{X}}$ be a probability vector that we know how to sample on the same set $\mathbb{X}$, and let $C \geq 1$ be such that $p_x \leq C q_x$ for all $x \in \mathbb{X}$. The classical von Neumann rejection algorithm is shown as Algorithm 1. It is well known that the expected number of times round the **repeat** loop is exactly C.

Algorithm 1 Original von Neumann rejection algorithm

1: **repeat**
2:   Sample X according to $(q_x)_{x \in \mathbb{X}}$
3:   Sample U uniformly on $[0,1]$
4:   **if** $UCq_x \leq p_x$ **then**
5:     **return** X {X is accepted}
6:   **end if**
7: **end repeat**

If only partial knowledge about the parameters defining p is known, it would seem that the condition in line 4 cannot be decided. Nevertheless, the strategy is to build a sequence of increasingly accurate approximations that converge to the left and right sides of the test. As explained below, the number of bits transmitted depends on the number of bits needed to compute q, and on the accuracy in p required to accept or reject. This task can be achieved either in the *random bit model*, in which only i.i.d. random bits are generated, or in the less realistic *uniform model*, in which uniform continuous random variables are needed. The random bit model was originally suggested by von Neumann [18], but only later given this name and formalized by Knuth and Yao [21]. In this section, we concentrate for simplicity on the uniform model, leaving the more practical random bit model for Section 4.

Definition 1. *A t-bit approximation of a real number x is any $\hat{x}$ such that $|x - \hat{x}| \leq 2^{-t}$. A special case of t-bit approximation is the t-bit truncation $\hat{x} = \text{sign}(x) \lfloor |x| 2^t \rfloor / 2^t$, where $\text{sign}(x)$ is equal to $+1$, 0 or -1 depending on the sign of x. If $\alpha = a + bi$ is a complex number, where $i = \sqrt{-1}$, then a t-bit approximation (resp. truncation) $\hat{\alpha}$ of α is any $\hat{a} + \hat{b}i$, where $\hat{a}$ and $\hat{b}$ are t-bit approximations (resp. truncations) of a and b, respectively.*

Note that we assume without loss of generality that approximations of probabilities are always constrained to be real numbers between 0 and 1, which can be enforced by snapping any out-of-bound approximation (even if it is a complex number) to the closest valid value.

Consider an integer $t_0 > 0$ to be determined later. Our strategy is for the leader to compute the probability vector $q = (q_x)_{x \in \mathbb{X}}$ defined below, based on t_0-bit approximations $p_x(t_0)$ of the probabilities p_x for each $x \in \mathbb{X}$. For this purpose, the leader receives sufficient information from the custodians to build the entire vector q at the outset of the protocol. This makes q the "easy-to-sample" distribution required in von Neumann's technique, which is easy not from a computational viewpoint, but in the sense that no further communication is required for the leader to sample it as many times as needed.

Let
$$C = \sum_x \left(p_x(t_0) + 2^{-t_0}\right) \tag{1}$$

and
$$q_x = \left(p_x(t_0) + 2^{-t_0}\right)/C. \tag{2}$$

Noting that $\sum_x q_x = 1$, these q_x define a proper probability vector $q = (q_x)_{x \in \mathbb{X}}$. Using the definition of a t-bit approximation and the definition of q_x from Equation (2), we have that

$$p_x \leq \left(p_x(t_0) + 2^{-t_0} = Cq_x\right) \leq p_x + 2 \times 2^{-t_0}.$$

Taking the sum over the possible values for x and recalling that set $\mathbb{X}$ is of cardinality n,

$$1 \leq C \leq 1 + 2^{1-t_0} n. \tag{3}$$

Consider any $x \in \mathbb{X}$ sampled according to q and U sampled uniformly in $[0,1]$ as in lines 2 and 3 of Algorithm 1. Should x be accepted because $UCq_x \leq p_x$, this can be certified by any t-bit approximation $p_x(t)$ of p_x such that $UCq_x \leq p_x(t) - 2^{-t}$ for some positive integer t since $p_x(t) \leq p_x + 2^{-t}$. Conversely, any integer t such that $UCq_x > p_x(t) + 2^{-t}$ certifies that x should be rejected because it implies that $UCq_x > p_x$ since $p_x(t) \geq p_x - 2^{-t}$. On the other hand, no decision can be made concerning UCq_x versus p_x if $-2^{-t} < UCq_x - p_x(t) \leq 2^{-t}$. It follows that one can modify Algorithm 1 above into Algorithm 2 below, in which a sufficiently precise approximation of p_x suffices to make the correct decision to accept or reject an x sampled according to distribution q. A well-chosen value of t_0 must be input into this algorithm, as discussed later.

Algorithm 2 Modified rejection algorithm—Protocol for the leader
---
Input: Value of t_0
1: Compute $p_x(t_0)$ for each $x \in \mathbb{X}$
   {The leader needs information from the custodians in order to compute these approximations}
2: Compute C and $q = (q_x)_{x \in \mathbb{X}}$ as per Equations (1) and (2)
3: Sample X according to q
4: Sample U uniformly on $[0,1]$
5: **for** $t = t_0$ **to** ∞ **do**
6:    **if** $UCq_x \leq p_x(t) - 2^{-t}$ **then**
7:       **return** X {X is accepted}
8:    **else if** $UCq_x > p_x(t) + 2^{-t}$ **then**
9:       **go to** line 3 {X is rejected}
10:    **else**
11:       Continue the **for** loop
   {We cannot decide whether to accept or reject because $-2^{-t} < UCq_x - p_x(t) \leq 2^{-t}$; communication may be required in order for the leader to compute $p_x(t+1)$; it could be that bits previously communicated to compute $p_x(t)$ can be reused.}
12:    **end if**
13: **end for**

Theorem 1. *Algorithm 2 is correct, i.e., it terminates and returns $X = x$ with probability p_x. Furthermore, let T be the random variable that denotes the value of variable t upon termination of any instance of the **for** loop, whether the loop terminates in rejection or acceptance. Then,*

$$\mathbf{E}(T) \leq t_0 + 3. \tag{4}$$

Proof. Consider any $x \in \mathbb{X}$ and $t \geq t_0$. To reach $T > t$, it must be that $-2^{-t} < UCq_x - p_x(t) \leq 2^{-t}$. Noting that $q_x \neq 0$ according to Equation (2), the probability that $T > t$ when $X = x$ is therefore upper-bounded as follows:

$$\begin{aligned}
\mathbf{P}\{T > t \mid X = x\} &\leq \mathbf{P}\{-2^{-t} < UCq_x - p_x(t) \leq 2^{-t}\} \\
&= \mathbf{P}\left\{\frac{p_x(t) - 2^{-t}}{Cq_x} < U \leq \frac{p_x(t) + 2^{-t}}{Cq_x}\right\} \\
&\leq \frac{p_x(t) + 2^{-t}}{Cq_x} - \frac{p_x(t) - 2^{-t}}{Cq_x} = \frac{2 \times 2^{-t}}{Cq_x} \leq 2^{t_0 - t + 1}.
\end{aligned} \tag{5}$$

The last inequality uses the fact that

$$Cq_x = p_x(t_0) + 2^{-t_0} \geq 2^{-t_0}.$$

It follows that the probability that more turns round the **for** loop are required decreases exponentially with each new turn once $t > t_0 + 1$, which suffices to guarantee termination of the **for** loop with probability 1. Termination of the algorithm itself comes from the fact that the choice of X and U in lines 3 and 4 leads to acceptance at line 7—and therefore termination—with probability $1/C$, as demonstrated by von Neumann in the analysis of his rejection algorithm.

The fact that $X = x$ is returned with probability p_x is an immediate consequence of the correctness of the von Neumann rejection algorithm since our adaptation of this method to handle the fact that only approximations of p_x are available does not change the decision to accept or reject any given candidate sampled according to q.

In order to bound the expectation of T, we note that $\mathbf{P}\{T > t \mid X = x\} = 1$ when $t < t_0$ since we start the **for** loop at $t = t_0$. We can also use vacuous $\mathbf{P}\{T > t_0 \mid X = x\} \leq 1$ rather than the worse-than-vacuous upper bound of 2 given by Equation (5) in the case $t = t_0$. Therefore,

$$\begin{aligned}
\mathbf{E}(T \mid X = x) &= \sum_{t=0}^{\infty} \mathbf{P}\{T > t \mid X = x\} \\
&= \sum_{t=0}^{t_0} \mathbf{P}\{T > t \mid X = x\} + \sum_{t=t_0+1}^{\infty} \mathbf{P}\{T > t \mid X = x\} \\
&\leq t_0 + 1 + 2^{t_0+1} \sum_{t=t_0+1}^{\infty} 2^{-t} = t_0 + 3.
\end{aligned}$$

It remains to note that, since $\mathbf{E}(T \mid X = x) \leq t_0 + 3$ for all $x \in \mathbb{X}$, it follows that $\mathbf{E}(T) \leq t_0 + 3$ without condition. □

Let S be the random variable that represents the number of times variable X is sampled according to q at line 3, and let T_i be the random variable that represents the value of variable T upon termination of the ith instance of the **for** loop starting at line 5, for $i \in \{1, \ldots, S\}$. The random variables T_i are independently and identically distributed as the random variable T in Theorem 1 and the expected value of S is C. Let $X_1, \ldots, X_S$ be the random variables chosen at successive passes at line 3, so that $X_1, \ldots, X_{S-1}$ are rejected, whereas X_S is returned as the final result of the algorithm.

To analyse the communication complexity of Algorithm 2, we introduce function $\gamma_x(t)$ for each $x \in \mathbb{X}$ and $t > t_0$, which denotes the *incremental* number of bits that the leader must receive from the custodians in order to compute $p_x(t)$, taking account of the information that may already be available if he had previously computed $p_x(t-1)$. For completeness, we include in $\gamma_x(t)$ the cost of the communication required for the leader to request more information from the custodians. We also introduce function $\delta(t)$ for $t \geq 0$, which denotes the number of bits that the leader must receive from the custodians in order to compute $p_x(t)$ for all $x \in \mathbb{X}$ in a "simultaneous" manner. Note that it could be much less expensive to compute those n values than n times the cost of computing any single one of them because some of the parameters held by the custodians may be relevant to more than one of the p_x's. The total number of bits communicated in order to implement Algorithm 2 is therefore given by random variable

$$Z = \delta(t_0) + \sum_{i=1}^{S} \sum_{t=t_0+1}^{T_i} \gamma_{x_i}(t).$$

For simplicity, let us define function $\gamma(t) \stackrel{\text{def}}{=} \max_{x \in \mathbb{X}} \gamma_x(t)$. We then have

$$Z \leq \delta(t_0) + \sum_{i=1}^{S} \sum_{t=t_0+1}^{T_i} \gamma(t),$$

whose expectation, according to Wald's identity, is

$$\mathbf{E}(Z) \leq \delta(t_0) + \mathbf{E}(S)\, \mathbf{E}\left(\sum_{t=t_0+1}^{T} \gamma(t) \right). \qquad (6)$$

Assuming the value of $\gamma(t)$ is upper-bounded by some γ,

$$\begin{aligned}
\mathbf{E}(Z) &\leq \delta(t_0) + \mathbf{E}(S)\mathbf{E}(T-t_0)\gamma \\
&\leq \delta(t_0) + 3\gamma C \\
&\leq \delta(t_0) + 3\gamma(1 + 2^{1-t_0}n)
\end{aligned} \qquad (7)$$

because $\mathbf{E}(S) = C$ and using Equations (4) and (3).

Depending on the specific application, which determines γ and function $\delta(t)$, Equation (7) is key to a trade-off that can lead to an optimal choice of t_0 since a larger t_0 decreases 2^{1-t_0} but is likely to increase $\delta(t_0)$. The value of γ may play a rôle in the balance. The next section, in which we consider the simulation of quantum entanglement by classical communication, gives an example of this trade-off in action.

3. Simulation of Quantum Entanglement Based on Remote Sampling

Before introducing the simulation of entanglement, let us establish some notation and mention the mathematical objects that we shall need. It is assumed that the reader is familiar with linear algebra, in particular the notion of a semi-definite positive matrix, Hermitian matrix, trace of a matrix, tensor product, etc. For a discussion about the probabilistic and statistical nature of quantum theory, see Ref. [22]. For convenience, we use $[n]$ to denote the set $\{1, 2, \ldots, n\}$ for any integer n.

Consider integers $m, d_1, d_2, \ldots, d_m, n_1, n_2, \ldots, n_m$, all greater than or equal to 2. Define $d = \prod_{i=1}^{m} d_i$ and $n = \prod_{i=1}^{m} n_i$. Let ρ be a $d \times d$ density matrix. Recall that any density matrix is Hermitian, semi-definite positive and unit-trace, which implies that its diagonal elements are real numbers between 0 and 1. For each $i \in [m]$ and $j \in [n_i]$, let M_{ij} be a $d_i \times d_i$ Hermitian semi-definite positive matrix such that

$$\sum_{j \in [n_i]} M_{ij} = I_{d_i}, \qquad (8)$$

where I_{d_i} is the $d_i \times d_i$ identity matrix. In other words, each set $\{M_{ij}\}_{j \in [n_i]}$ is a POVM (positive-operator valued measure) [22].

As introduced in Section 1, we consider one *leader* and *m custodians*. The leader knows density matrix ρ and the ith custodian knows the ith POVM, meaning that he knows the matrices M_{ij} for all $j \in [n_i]$. If a physical system of dimension d in state ρ were shared between the custodians, in the sense that the ith custodian had possession of the ith subsystem of dimension d_i, each custodian could perform locally his assigned POVM and output the outcome, an integer between 1 and n_i. The joint output would belong to $\mathbb{X} \stackrel{\text{def}}{=} [n_1] \times [n_2] \times \cdots \times [n_m]$, a set of cardinality n, sampled according to the probability distribution stipulated by the laws of quantum theory, which we review below.

Our task is to sample $\mathbb{X}$ with the exact same probability distribution even though there is no physical system in state ρ available to the custodians, and in fact all parties considered are purely classical! We know from Bell's Theorem [23] that this task is impossible in general without communication, even when $m = 2$, and our goal is to minimize the amount of communication required to achieve it. Special cases of this problem have been studied extensively for expected [1,2,4–6], etc. and worst-case [3,8], etc. communication complexity, but here we solve it in its essentially most general setting, albeit only in the expected sense. For this purpose, the leader will centralize the operations while requesting as little information as possible from the custodians on their assigned POVMs. Once the leader has successfully sampled $X = (X_1, \ldots, X_m)$, he transmits each X_i to the ith custodian, who can then output it as would have been the case had quantum measurements actually taken place.

We now review the probability distribution $\mathbb{X}$ that we need to sample, according to quantum theory. For each vector $x = (x_1, \ldots, x_m) \in \mathbb{X}$, let M_x be the $d \times d$ tensor product of matrices M_{ix_i} for each $i \in [m]$:

$$M_x = \bigotimes_{i=1}^{m} M_{ix_i}. \qquad (9)$$

The set $\{M_x\}_{x \in \mathbb{X}}$ forms a global POVM of dimension d, which applied to density matrix ρ defines a joint probability vector on $\mathbb{X}$. The probability p_x of obtaining any $x = (x_1, \ldots, x_m) \in \mathbb{X}$ is given by

$$p_x = \text{Tr}(\rho M_x) = \text{Tr}\left(\rho \left(\bigotimes_{i=1}^{m} M_{ix_i}\right)\right). \qquad (10)$$

For a matrix A of size $s \times s$ and any pair of indices r and c between 0 and $s - 1$, we use $(A)_{rc}$ to denote the entry of A located in the rth row and c^{th} column. Matrix indices start at 0 rather than 1 to facilitate Fact 2 below. We now state various facts for which we provide cursory justifications since they follow from elementary linear algebra and quantum theory, or they are lifted from previous work.

Fact 1. For all $x \in \mathbb{X}$, we have $0 \leq p_x \leq 1$ when p_x is defined according to Equation (10); furthermore, $\sum_{x \in \mathbb{X}} p_x = 1$. This is obvious because quantum theory tells us that Equation (10) defines a probability distribution over all possible outcomes $x \in \mathbb{X}$, as sampled by the joint measurement. Naturally, this statement could also be proven from Equations (8) and (10) using elementary linear algebra.

Fact 2. For each $x = (x_1, \ldots, x_m) \in \mathbb{X}$, matrix M_x is the tensor product of m matrices as given in Equation (9). Therefore, each entry $(M_x)_{rc}$ is the product of m entries of the M_{ix_i}'s. Specifically, consider any indices r and c between 0 and $d - 1$ and let r_i and c_i be the indices between 0 and $d_i - 1$, for each $i \in [m]$, such that

$$r = r_1 + r_2 d_1 + r_3 d_1 d_2 + \ldots + r_m d_1 \cdots d_{m-1},$$
$$c = c_1 + c_2 d_1 + c_3 d_1 d_2 + \ldots + c_m d_1 \cdots d_{m-1}.$$

The r_i's and c_i's are uniquely defined by the principle of mixed-radix numeration. We have

$$(M_x)_{rc} = \prod_{i=1}^{m} (M_{ix_i})_{r_i c_i}.$$

Fact 3. Let M be a Hermitian semi-definite positive matrix. Every entry $(M)_{ij}$ of the matrix satisfies

$$|(M)_{ij}| \leq \sqrt{(M)_{ii}(M)_{jj}}.$$

This follows from the fact that all principal submatrices of any Hermitian semi-definite positive matrix are semi-definite positive [24] (Observation 7.1.2, page 430). In particular, the principal submatrix

$$\begin{pmatrix} (M)_{ii} & (M)_{ij} \\ (M)_{ji} & (M)_{jj} \end{pmatrix}$$

is semi-definite positive, and therefore it has nonnegative determinant:

$$(M)_{ii}(M)_{jj} - (M)_{ij}(M)_{ji} = (M)_{ii}(M)_{jj} - (M)_{ij}(M)_{ij}^* = (M)_{ii}(M)_{jj} - |(M)_{ij}|^2 \geq 0$$

by virtue of M being Hermitian, where α^* denotes the complex conjugate of α.

Fact 4. The norm $|(\rho)_{ij}|$ of any entry of a density matrix ρ is less than or equal to 1. This follows directly from Fact 3 since density matrices are Hermitian semi-definite positive, and from the fact that diagonal entries of density matrices, such as $(\rho)_{ii}$ and $(\rho)_{jj}$, are real values between 0 and 1.

Fact 5. Given any POVM $\{M_\ell\}_{\ell=1}^L$, we have that

1. $0 \leq (M_\ell)_{ii} \leq 1$ for all ℓ and i, and
2. $|(M_\ell)_{ij}| \leq 1$ for all ℓ, i and j.

The first statement follows from the fact that $\sum_{\ell=1}^L M_\ell$ is the identity matrix by definition of POVMs, and therefore $\sum_{\ell=1}^L (M_\ell)_{ii} = 1$ for all i, and the fact that each $(M_\ell)_{ii} \geq 0$ because each M_ℓ is semi-definite positive. The second statement follows from the first by applying Fact 3.

Fact 6 (This is a special case of Theorem 1 from Ref. [1], with $v = 0$). Let $k \geq 1$ be an integer and consider any two real numbers a and b. If $\hat{a}$ and $\hat{b}$ are arbitrary k-bit approximations of a and b, respectively, then $\hat{a} + \hat{b}$ is a $(k-1)$-bit approximation of $a + b$. If, in addition, a and b are known to lie in interval $[-1, 1]$, which can also be assumed without loss of generality concerning $\hat{a}$ and $\hat{b}$ since otherwise they can be safely pushed back to the appropriate frontier of this interval, then $\hat{a}\hat{b}$ is a $(k-1)$-bit approximation of ab.

Fact 7. Let $k \geq 1$ be an integer and consider any two *complex* numbers α and β. If $\hat{\alpha}$ and $\hat{\beta}$ are arbitrary k-bit approximations of α and β, respectively, then $\hat{\alpha} + \hat{\beta}$ is a $(k-1)$-bit approximation of $\alpha + \beta$. If, in addition, $k \geq 2$ and the real and imaginary parts of α and β are known to lie in interval $[-1, 1]$, which can also be assumed without loss of generality concerning $\hat{\alpha}$ and $\hat{\beta}$, then $\hat{\alpha}\hat{\beta}$ is a $(k-2)$-bit approximation of $\alpha\beta$. This is a direct consequence of Fact 6 in the case of addition. In the case of multiplication, consider $\alpha = a + bi$, $\beta = c + di$, $\hat{\alpha} = \hat{a} + \hat{b}i$ and $\hat{\beta} = \hat{c} + \hat{d}i$, so that

$$\alpha\beta = (ac - bd) + (ad + bc)i \quad \text{and} \quad \hat{\alpha}\hat{\beta} = (\hat{a}\hat{c} - \hat{b}\hat{d}) + (\hat{a}\hat{d} + \hat{b}\hat{c})i.$$

By the multiplicative part of Fact 6, $\hat{a}\hat{c}$, $\hat{b}\hat{d}$, $\hat{a}\hat{d}$ and $\hat{b}\hat{c}$ are $(k-1)$-bit approximations of ac, bd, ad and bc, respectively; and then by the additive part of the same fact (which obviously applies equally well to subtraction), $\hat{a}\hat{c} - \hat{b}\hat{d}$ and $\hat{a}\hat{d} + \hat{b}\hat{c}$ are $(k-2)$-bit approximations of $ac - bd$ and $ad + bc$, respectively.

Fact 8 (This is Corollary 2 from Ref. [1]). Let $m \geq 2$ and $k \geq \lceil \lg m \rceil$ be integers and let $\{a_j\}_{j=1}^m$ and $\{\hat{a}_j\}_{j=1}^m$ be real numbers and their k-bit approximations, respectively, all in interval $[-1, 1]$. Then, $\prod_{j=1}^m \hat{a}_j$ is a $(k - \lceil \lg m \rceil)$-bit approximation of $\prod_{j=1}^m a_j$.

Fact 9. Let $m \geq 2$ and $k \geq 2\lceil \lg m \rceil$ be integers and let $\{\alpha_j\}_{j=1}^m$ and $\{\hat{\alpha}_j\}_{j=1}^m$ be complex numbers and their k-bit approximations, respectively. Provided it is known that $|\alpha_j| \leq 1$ for each $j \in [m]$, a $(k - 2\lceil \lg m \rceil)$-bit approximation of $\prod_{j=1}^m \alpha_j$ can be computed from knowledge of the $\hat{\alpha}_j$'s. The proof of this fact follows essentially the same template as Fact 8, except that *two* bits of precision may be lost at each level up the binary tree introduced in Ref. [1], due to the difference between Facts 6 and 7. A subtlety occurs in the need for Fact 7 to apply that the real and imaginary parts of all the complex numbers under consideration must lie in interval $[-1, 1]$. This is automatic for the exact values since the α_j's are upper-bounded in norm by 1 and the product of such-bounded complex numbers is also upper-bounded in norm by 1, which implies that their real and imaginary parts lie in interval $[-1, 1]$. For the approximations, however, we cannot force their *norm* to be bounded by 1 because we need the approximations to be rational for communication purposes. Fortunately, we can force the real and imaginary parts of all approximations computed at each level up the binary tree to lie in interval $[-1, 1]$ because we know that they approximate such-bounded numbers. Note that the product of two complex numbers whose real and imaginary parts lie in interval $[-1, 1]$, such as $1 + 2^{-k}i$ and $1 - 2^{-k}i$, may not have this property, even if they are k-bit approximations of numbers bounded in norm by 1.

Fact 10. Let $s \geq 2$ and $k \geq \lceil \lg s \rceil$ be integers and let $\{\alpha_j\}_{j=1}^s$ and $\{\hat{\alpha}_j\}_{j=1}^s$ be complex numbers and their k-bit approximations, respectively, without any restriction on their norm. Then $\sum_{j=1}^s \hat{\alpha}_j$ is a $(k - \lceil \lg s \rceil)$-bit approximation of $\sum_{j=1}^s \alpha_j$. Again, this follows the same proof template as Fact 8, substituting multiplication of real numbers by addition of complex numbers, which allows us to drop any condition on the size of the numbers considered.

Fact 11. Consider any $x = (x_1, \ldots, x_m) \in \mathbb{X}$ and any positive integer t. In order to compute a t-bit approximation of p_x, it suffices to have $(t + 1 + \lceil 2 \lg d \rceil + 2 \lceil \lg m \rceil)$-bit approximations of each entry of the M_{ix_i} matrices for all $i \in [m]$. This is because

$$p_x = \mathrm{Tr}(\rho M_x) = \sum_{r=0}^{d-1} (\rho M_x)_{rr}$$

$$= \sum_{r=0}^{d-1} \sum_{c=0}^{d-1} (\rho)_{rc} (M_x)_{cr}$$

$$= \sum_{r=0}^{d-1} \sum_{c=0}^{d-1} (\rho)_{rc} \prod_{i=1}^{m} (M_{ix_i})_{c_i r_i} \tag{11}$$

by virtue of Fact 2. Every term of the double sum in Equation (11) involves a product of m entries, one per POVM element, and therefore incurs a loss of at most $2 \lceil \lg m \rceil$ bits of precision by Fact 9, whose condition holds thanks to Fact 5. An additional bit of precision may be lost in the multiplication by $(\rho)_{rc}$, even though that value is available with arbitrary precision (and is upper-bounded by 1 in norm by Fact 4) because of the additions involved in multiplying complex numbers. Then, we have to add $s = d^2$ terms, which incurs an additional loss of at most $\lceil \lg s \rceil = \lceil 2 \lg d \rceil$ bits of precision by Fact 10. In total, $(t + 1 + \lceil 2 \lg d \rceil + 2 \lceil \lg m \rceil)$-bit approximations of the $(M_{ix_i})_{c_i r_i}$'s will result in a t-bit approximation of p_x.

Fact 12. The leader can compute $p_x(t)$ for any specific $x = (x_1, \ldots, x_m) \in \mathbb{X}$ and integer t if he receives a total of

$$(t + 2 + \lceil 2 \lg d \rceil + 2 \lceil \lg m \rceil) \sum_{i=1}^{m} d_i^2$$

bits from the custodians. This is because the ith custodian has the description of matrix M_{ix_i} of size $d_i \times d_i$, which is defined by exactly d_i^2 real numbers since the matrix is Hermitian. By virtue of Fact 11, it is sufficient for the leader to have $(t + 1 + \lceil 2 \lg d \rceil + 2 \lceil \lg m \rceil)$-bit approximations for all those $\sum_{i=1}^{m} d_i^2$ numbers. Since each one of them lies in interval $[-1, 1]$ by Fact 5, well-chosen k-bit approximations (for instance k-bit truncations) can be conveyed by the transmission of $k + 1$ bits, one of which carries the sign.

Note that the t-bit approximation of p_x computed according to Fact 12, say $a + bi$, may very well have a nonzero imaginary part b, albeit necessarily between -2^{-t} and 2^{-t}. Since $p_x(t)$ must be a real number between 0 and 1, the leader sets $p_x(t) = \max(0, \min(1, a))$, taking no account of b, although a paranoid leader may wish to test that $-2^{-t} \leq b \leq 2^{-t}$ indeed and raise an alarm in case it is not (which of course is mathematically impossible unless the custodians are not given proper POVMs, unless they misbehave, or unless a computation or communication error has occurred).

Fact 13. For any t, the leader can compute $p_x(t)$ for each and every $x \in \mathbb{X}$ if he receives

$$\delta(t) \stackrel{\text{def}}{=} (t + 2 + \lceil 2\lg d \rceil + 2\lceil \lg m \rceil) \sum_{i=1}^{m} n_i d_i^2$$

bits from the custodians. This is because it suffices for each custodian i to send to the leader $(t + 1 + \lceil 2\lg d \rceil + 2\lceil \lg m \rceil)$-bit approximations of all $n_i d_i^2$ real numbers that define the entire ith POVM, i.e., all the matrices M_{ij} for $j \in [n_i]$. This is a nice example of the fact that it may be much less expensive for the leader to compute at once $p_x(t)$ for all $x \in \mathbb{X}$, rather than computing them one by one independently, which would cost

$$n(t + 2 + \lceil 2\lg d \rceil + 2\lceil \lg m \rceil) \sum_{i=1}^{m} d_i^2 = (t + 2 + \lceil 2\lg d \rceil + 2\lceil \lg m \rceil) \sum_{i=1}^{m} nd_i^2 \gg \delta(t)$$

bits of communication by applying n times Fact 12.

After all these preliminaries, we are now ready to adapt the general template of Algorithm 2 to our entanglement-simulation conundrum, yielding Algorithm 3. We postpone the choice of t_0 until after the communication complexity analysis of this new algorithm.

Algorithm 3 Protocol for simulating arbitrary entanglement subjected to arbitrary measurements

1: Each custodian $i \in [m]$ sends his value of n_i to the leader, who computes $n = \prod_{i=1}^{m} n_i$
2: The leader chooses t_0 and informs the custodians of its value
3: Each custodian $i \in [m]$ sends to the leader $(t_0 + 1 + \lceil 2\lg d \rceil + 2\lceil \lg m \rceil)$-bit truncations of the real and imaginary parts of the entries defining matrix M_{ij} for each $j \in [n_i]$
4: The leader computes $p_x(t_0)$ for every $x \in \mathbb{X}$, using Fact 13
5: The leader computes C and $q = (q_x)_{x \in \mathbb{X}}$ as per Equations (1) and (2)
6: accept ← false
7: **repeat**
8:     reject ← false
9:     The leader samples $X = (X_1, X_2, \ldots, X_m)$ according to q
10:     The leader informs each custodian $i \in [m]$ of the value of X_i
11:     The leader samples U uniformly on $[0, 1]$
12:     $t \leftarrow t_0$
13:     **repeat**
14:         **if** $UCq_x \leq p_x(t) - 2^{-t}$ **then**
15:             accept ← true {X is accepted}
16:         **else if** $UCq_x > p_x(t) + 2^{-t}$ **then**
17:             reject ← true {X is rejected}
18:         **else**
19:             The leader asks each custodian $i \in [m]$ for one more bit in the truncation of the real and imaginary parts of the entries defining matrix M_{iX_i};
20:             Using this information, the leader updates $p_x(t)$ into $p_x(t+1)$;
21:             $t \leftarrow t + 1$
22:         **end if**
23:     **until** accept **or** reject
24: **until** accept
25: The leader requests each custodian $i \in [m]$ to output his X_i

To analyse the expected number of bits of communication required by this algorithm, we apply Equation (7) from Section 2 after defining explicitly the cost parameters $\delta(t_0)$ for the initial computation of $p_x(t_0)$ for all $x \in \mathbb{X}$ at lines 3 and 4, and γ for the upgrade from a specific $p_x(t)$ to $p_x(t+1)$ at lines 19 and 20. For simplicity, we shall ignore the negligible amount of communication entailed at line 1 (which is $\sum_{i=1}^{m} \lceil \lg n_i \rceil \leq m + \lg n$ bits), line 2 ($\lceil \lg t_0 \rceil$ bits), line 10 (also $\sum_{i=1}^{m} \lceil \lg n_i \rceil$ bits, but repeated $\mathbf{E}(S) \leq 1 + 2^{1-t_0} n$ times) and line 25 (m bits) because they are not taken into account in Equation (7) since they are absent from Algorithm 2. If we counted it all, this would add $O((1 + 2^{1-t_0} n) \lg n + \lg t_0)$ bits to Equation (13) below, which would be less than $10 \lg n$ bits added to Equation (14), with no effect at all on Equation (15).

According to Fact 13,

$$\delta(t_0) = (t_0 + 2 + \lceil 2 \lg d \rceil + 2 \lceil \lg m \rceil) \sum_{i=1}^{m} n_i d_i^2.$$

The cost of line 19 is very modest because we use *truncations* rather than general approximations in line 3 for the leader to compute $p_x(t_0)$ for all $x \in \mathbb{X}$. Indeed, it suffices to obtain a single additional bit of precision in the real and imaginary parts of each entry defining matrix M_{ix_i} from each custodian $i \in [m]$. The cost of this update is simply

$$\gamma = m + \sum_{i=1}^{m} d_i^2 \qquad (12)$$

bits of communication, where the addition of m is to account for the leader needing to request new bits from the custodians. This is a nice example of what we meant by "it could be that bits previously communicated can be reused" in line 11 of Algorithm 2.

Putting it all together in Equation (7), the total expected number of bits communicated in Algorithm 3 in order to sample exactly according to the quantum probability distribution is

$$\mathbf{E}(Z) \leq \delta(t_0) + 3\gamma(1 + 2^{1-t_0} n)$$
$$\leq (t_0 + 2 + \lceil 2 \lg d \rceil + 2 \lceil \lg m \rceil) \sum_{i=1}^{m} n_i d_i^2 + 3(1 + 2^{1-t_0} n)\left(m + \sum_{i=1}^{m} d_i^2\right). \qquad (13)$$

We are finally in a position to choose the value of parameter t_0. First note that $n = \prod_{i=1}^{m} n_i \geq 2^m$. Therefore, any constant choice of t_0 will entail an expected amount of communication that is exponential in m because of the right-hand term in Equation (13). At the other extreme, choosing $t_0 = n$ would also entail an expected amount of communication that is exponential in m, this time because of the left-hand term in Equation (13). A good compromise is to choose $t_0 = \lceil \lg n \rceil$, which results in $1 \leq C \leq 3$ according to Equation (3), because in that case $2^{t_0} \geq n$ and therefore

$$1 \leq C \leq 1 + 2^{1-t_0} n = 1 + \frac{2n}{2^{t_0}} \leq 3,$$

so that Equation (13) becomes

$$\mathbf{E}(Z) \leq (\lceil \lg n \rceil + \lceil 2 \lg d \rceil + 2 \lceil \lg m \rceil + 2) \sum_{i=1}^{m} n_i d_i^2 + 9\left(m + \sum_{i=1}^{m} d_i^2\right). \qquad (14)$$

In case all the n_i's and d_i's are upper-bounded by some constant ζ, we have that $n = \prod_{i=1}^{m} n_i \leq \zeta^m$, hence $\lg n \leq m \lg \zeta$, similarly $\lg d \leq m \lg \zeta$, and also $\sum_{i=1}^{m} n_i d_i^2 \leq m \zeta^3$. It follows that

$$\mathbf{E}(Z) \leq (3 \zeta^3 \lg \zeta) m^2 + O(m \log m), \qquad (15)$$

which is on the order of m^2, thus matching with our most general method the result that was already known for the very specific case of simulating the quantum m-partite GHZ distribution [1].

4. Practical Implementation Using a Source of Discrete Randomness

In practice, we cannot work with continuous random variables since our computers have finite storage capacities and finite precision arithmetic. Furthermore, the generation of uniform continuous random variables does not make sense computationally speaking and we must adapt Algorithms 2 and 3 to work in a finite world.

For this purpose, recall that U is a uniform continuous random variable on $[0,1]$ used in all the algorithms seen so far. For each $i \geq 1$, let U_i denote the ith bit in the binary expansion of U, so that

$$U = 0.U_1 U_2 \cdots = \sum_{i=1}^{\infty} U_i 2^{-i}.$$

We acknowledge the fact that the U_i's are not uniquely defined in case $U = j/2^k$ for integers $k > 0$ and $0 < j < 2^k$, but we only mention this phenomenon to ignore it since it occurs with probability 0 when U is uniformly distributed on $[0,1]$. We denote the t-bit truncation of U by $U[t]$:

$$U[t] \stackrel{\text{def}}{=} \lfloor 2^t U \rfloor / 2^t = \sum_{i=1}^{t} U_i 2^{-i}.$$

For all $t \geq 1$, we have that

$$U[t] \leq U < U[t] + 2^{-t}. \tag{16}$$

We modify Algorithm 2 into Algorithm 4 as follows, leaving to the reader the corresponding modification of Algorithm 3, thus yielding a practical protocol for the simulation of general entanglement under arbitrary measurements.

Algorithm 4 Modified rejection algorithm with discrete randomness source—Protocol for the leader

---

Input: Value of t_0
1: Compute $p_x(t_0)$ for each $x \in \mathbb{X}$
   {The leader needs information from the custodians in order to compute these approximations}
2: Compute C and $q = (q_x)_{x \in \mathbb{X}}$ as per Equations (1) and (2)
3: Sample X according to q
4: $U[0] \leftarrow 0$
5: **for** $t = 1$ **to** $t_0 - 1$ **do**
6:    Generate i.i.d. unbiased bit U_t
7:    $U[t] \leftarrow U[t-1] + U_t 2^{-t}$
8: **end for**
9: **for** $t = t_0$ **to** ∞ **do**
10:   Generate i.i.d. unbiased bit U_t
11:   $U[t] \leftarrow U[t-1] + U_t 2^{-t}$
12:   **if** $(U[t] + 2^{-t})Cq_x \leq p_x(t) - 2^{-t}$ **then**
13:     **return** X {X is accepted}
14:   **else if** $U[t]Cq_x > p_x(t) + 2^{-t}$ **then**
15:     **go to line** 3 {X is rejected}
16:   **else**
17:     Continue the **for** loop
       {We cannot decide to accept or reject because $-(1+Cq_x)2^{-t} < U[t]Cq_x - p_x(t) \leq 2^{-t}$; communication may be required in order for the leader to compute $p_x(t+1)$; it could be that bits previously communicated to compute $p_x(t)$ can be reused.}
18:   **end if**
19: **end for**

Theorem 2. *Algorithm 4 is correct, i.e., it terminates and returns $X = x$ with probability p_x. Furthermore, let T be the random variable that denotes the value of variable t upon termination of any instance of the **for** loop that starts at line 9, whether it terminates in rejection or acceptance. Then,*

$$\mathbf{E}(T) \leq t_0 + 3 + 2^{-t_0}.$$

Proof. This is very similar to the proof of Theorem 1, so let us concentrate on the differences. First note that it follows from Equation (16) and the fact that $|p_x(t) - p_x| \leq 2^{-t}$ that

$$(U[t] + 2^{-t})Cq_x \leq p_x(t) - 2^{-t} \implies UCq_x \leq p_x(t) - 2^{-t} \implies UCq_x \leq p_x$$

and

$$U[t]Cq_x > p_x(t) + 2^{-t} \implies UCq_x > p_x(t) + 2^{-t} \implies UCq_x > p_x.$$

Therefore, whenever X is accepted at line 13 (resp. rejected at line 15), it would also have been accepted (resp. rejected) in the original von Neumann algorithm, which shows sampling correctness. Conversely, whenever we reach a value of $t \geq t_0$ such that $(U[t] + 2^{-t})Cq_x > p_x(t) - 2^{-t}$ and $U[t]Cq_x \leq p_x(t) + 2^{-t}$, we do not have enough information to decide whether to accept or reject, and therefore we reach line 17, causing t to increase. This happens precisely when

$$-(1 + Cq_x)2^{-t} < U[t]Cq_x - p_x(t) \leq 2^{-t}.$$

To obtain an upper bound on $\mathbf{E}(T)$, we mimic the proof of Theorem 1, but in the discrete rather than continuous regime. In particular, for any $x \in \mathbb{X}$ and $t \geq t_0$,

$$\mathbf{P}\{T > t \mid X = x\} \leq \mathbf{P}\{-(1 + Cq_x)2^{-t} < U[t]Cq_x - p_x(t) \leq 2^{-t}\}$$
$$= \mathbf{P}\{p_x(t) - (1 + Cq_x)2^{-t} < U[t]Cq_x \leq p_x(t) + 2^{-t}\}$$
$$= \mathbf{P}\left\{\frac{2^t p_x(t)}{Cq_x} - \frac{1 + Cq_x}{Cq_x} < 2^t U[t] \leq \frac{2^t p_x(t)}{Cq_x} + \frac{1}{Cq_x}\right\}$$
$$\leq \left[\left(\frac{2^t p_x(t)}{Cq_x} + \frac{1}{Cq_x}\right) - \left(\frac{2^t p_x(t)}{Cq_x} - \frac{1 + Cq_x}{Cq_x}\right) + 1\right]2^{-t} \quad (17)$$
$$= 2\left(1 + \frac{1}{Cq_x}\right)2^{-t} \leq 2^{t_0 - t + 1} + 2^{1-t} \quad \text{(because } Cq_x \geq 2^{-t_0}\text{)}. \quad (18)$$

To understand Equation (17), think of $2^t U[t]$ as an integer chosen randomly and uniformly between 0 and $2^t - 1$. The probability that it falls within some real interval $(a, b]$ for $a < b$ is equal to 2^{-t} times the number of integers between 0 and $2^t - 1$ in that interval, the latter being upper-bounded by the number of unrestricted integers in that interval, which is at most $b - a + 1$.

Noting how similar Equation (18) is to the corresponding Equation (5) in the analysis of Algorithm 2, it is not surprising that the expected value of T will be similar as well. Indeed, continuing as in the proof of Theorem 1, without belabouring the details,

$$\mathbf{E}(T \mid X = x) = \sum_{t=0}^{\infty} \mathbf{P}\{T > t \mid X = x\}$$
$$= \sum_{t=0}^{t_0+1} \mathbf{P}\{T > t \mid X = x\} + \sum_{t=t_0+2}^{\infty} \mathbf{P}\{T > t \mid X = x\}$$
$$\leq t_0 + 2 + 2^{t_0+1} \sum_{t=t_0+2}^{\infty} 2^{-t} + 2 \sum_{t=t_0+2}^{\infty} 2^{-t} = t_0 + 3 + 2^{-t_0}. \quad (19)$$

We conclude that $\mathbf{E}(T) \leq t_0 + 3 + 2^{-t_0}$ without condition since Equation (19) does not depend on x. □

The similarity between Theorems 1 and 2 means that there is no significant additional cost in the amount of communication required to achieve remote sampling in the random bit model. i.e., if we consider a realistic scenario in which the only source of randomness comes from i.i.d. unbiased bits, compared to an unrealistic scenario in which continuous random variables can be drawn. For instance, the reasoning that led to Equation (7) applies *mutatis mutandis* to conclude that the expected number Z of bits that needs to be communicated to achieve remote sampling in the random bit model is

$$\mathbf{E}(Z) \leq \delta(t_0) + (3 + 2^{-t_0})(1 + 2^{1-t_0}n)\gamma,$$

where δ and γ have the same meaning as in Section 2.

If we use the random bit approach for the general simulation of quantum entanglement studied in Section 3, choosing $t_0 = \lceil \lg n \rceil$ again, Equation (14) becomes

$$\mathbf{E}(Z) \leq (\lceil \lg n \rceil + \lceil 2 \lg d \rceil + 2\lceil \lg m \rceil + 2) \sum_{i=1}^{m} n_i d_i^2 + 3(3 + 1/n)\left(m + \sum_{i=1}^{m} d_i^2\right), \quad (20)$$

which reduces to the identical

$$\mathbf{E}(Z) \leq (3\zeta^3 \lg \zeta) m^2 + O(m \log m)$$

in case all the n_i's and d_i's are upper-bounded by some constant ζ, which again is on the order of m^2.

In addition to communication complexity, another natural efficiency measure in the random bit model concerns the *expected number of random bits* that needs to be drawn in order to achieve sampling. Randomness is needed in lines 3, 6 and 10 of Algorithm 4. A single random bit is required each time lines 6 and 10 are entered, but line 3 calls for sampling X according to distribution q. Let V_i be the random variable that represents the number of random bits needed on the ith passage through line 3. For this purpose, we use the algorithm introduced by Donald Knuth and Andrew Chi-Chih Yao [21], which enables sampling within any finite discrete probability distribution in the random bit model by using an expectation of no more than two random bits in addition to the Shannon binary entropy of the distribution. Since each such sampling is independent from the others, it follows that V_i is independently and identically distributed as a random variable V such that

$$\mathbf{E}(V) \leq 2 + H(q) \leq 2 + \lg n, \quad (21)$$

where $H(q)$ denotes the binary entropy of q, which is never more than the base-two logarithm of the number of atoms in the distribution, here n.

Let R be the random variable that represents the number of random bits drawn when running Algorithm 4. Reusing the notation of Section 2, let S be the random variable that represents the number of times variable X is sampled at line 3 and let T_i be the random variable that represents the value of variable T upon termination of the ith instance of the **for** loop starting at line 9, for $i \in \{1, \ldots, S\}$. The random variables T_i are independently and identically distributed as the random variable T in Theorem 2 and the expected value of S is C. Since one new random bit is generated precisely each time variable t is increased by 1 in any pass through either **for** loops (line 5 or 9), we simply have

$$R = \sum_{i=1}^{S} (V_i + T_i).$$

By virtue of Equations (3) and (21), Theorem 2, and using Wald's identity again, we conclude that:

$$\begin{aligned}\mathbf{E}(R) &= \mathbf{E}(S)\,(\mathbf{E}(V) + \mathbf{E}(T)) \\ &\leq (1 + 2^{1-t_0}n)(\lg n + t_0 + 5 + 2^{-t_0}).\end{aligned}$$

Taking $t_0 = \lceil \lg n \rceil$ again, remote sampling can be completed using an expected number of random bit in $O(\lg n)$, with a hidden multiplicative constant no larger than 6. The hidden constant can be reduced arbitrarily close to 2 by taking $t_0 = \lceil \lg n \rceil + a$ for an arbitrarily large constant a. Whenever target distribution p has close to full entropy, this is only twice the optimal number of random bits that would be required according to the Knuth–Yao lower bound [21] in the usual case when full knowledge of p is available in a central place rather than having to perform remote sampling. Note, however, that, if our primary consideration is to optimize communication for the classical simulation of entanglement, as in Section 3, choosing $t_0 = \lceil \lg n \rceil - a$ would be a better idea because the summation in the left-hand term of Equation (13) dominates that of the right-hand term. For this inconsequential optimization, a does not have to be a constant, but it should not exceed $\lg(\xi m)$, where ξ is our usual upper bound on the number of possible outcomes for each participant (if it exists), lest the right-hand term of Equation (13) overtake the left-hand term. Provided ξ exists, the expected number of random bits that needs to be drawn is linear in the number of participants.

The need for continuous random variables was not the only unrealistic assumption in Algorithms 1–3. We had also assumed implicitly that custodians know their private parameters precisely (and that the leader knows exactly each entry of density matrix ρ in Section 3). This could be reasonable in some situations, but it could also be that some of those parameters are transcendental numbers or the result of applying transcendental functions to other parameters, for example $\cos \pi/8$. More interestingly, it could be that the actual parameters are spoon-fed to the custodians by *examiners*, who want to test the custodians' ability to respond appropriately to unpredictable inputs. However, all we need is for the custodians to be able to obtain their own parameters with arbitrary precision, so that they can provide that information to the leader upon request. For example, if a parameter is $\pi/4$ and the leader requests a k-bit approximation of that parameter, the custodian can compute some number $\hat{x}$ such that $|\hat{x} - \pi/4| \leq 2^{-k}$ and provide it to the leader. For communication efficiency purposes, it is best if $\hat{x}$ itself requires only k bits to be communicated, or perhaps one more (for the sign) in case the parameter is constrained to be between -1 and 1. It is even better if the custodian can supply a k-bit *truncation* because this enables the possibility to upgrade it to a $(k+1)$-bit truncation by the transmission of a single bit upon request from the leader, as we have done explicitly for the simulation of entanglement at line 19 of Algorithm 3 in Section 3.

Nevertheless, it may be impossible for the custodians to compute truncations of their own parameters in some cases, even when they can compute arbitrarily precise approximations. Consider for instance a situation in which one parameter held by a custodian is $x = \cos \theta$ for some angle θ for which he can only obtain arbitrarily precise truncations. Unbeknownst to the custodian, $\theta = \pi/3$ and therefore $x = 1/2$. No matter how many bits the custodian obtains in the truncation of θ, however, he can never decide whether $\theta < \pi/3$ or $\theta \geq \pi/3$. In the first case, $x < 1/2$ and therefore the 1-bit truncation of x should be 0, whereas in the second (correct) case, $x \geq 1/2$ and therefore the 1-bit truncation of x is $1/2$ (or 0.1 in binary). Thus, the custodian will be unable to respond if the leader asks him for a 1-bit truncation of x, no matter how much time he spends on the task. In this example, by contrast, the custodian can supply the leader with arbitrarily precise *approximations* of x from appropriate approximations of θ. Should a situation like this occur, for instance in the simulation of entanglement, there would be two solutions. The first one is for the custodian to transmit increasingly precise truncations of θ to the leader and let *him* compute the cosine on it. This approach is only valid if it is known at the outset that the custodian's parameter will be of that form, which was essentially the solution taken in our earlier work on the simulation of the quantum m-partite GHZ distribution [1]. The more general solution is to modify the protocol and declare that custodians can send arbitrary approximations to the leader rather than truncations. The consequence on Algorithm 3 is that line 19 would become much more expensive since each custodian i would have to transmit a fresh one-bit-better approximation for the real and imaginary parts of the d_i^2 entries defining matrix M_{ix_i}. As a result, efficiency parameter $\gamma(t)$ in Equation (6) would become

$$\gamma(t) = m + (t + 2 + \lceil 2\lg d\rceil + 2\lceil \lg m\rceil)\sum_{i=1}^{m} d_i^2,$$

which should be compared with the much smaller (constant) value of γ given in Equation (12) when truncations of the parameters are available. Nevertheless, taking $t_0 = \lceil \lg n\rceil$ again, this increase in $\gamma(t)$ would make no significant difference in the total number of bits transmitted for the simulation of entanglement because it would increase only the right-hand term in Equations (14) and (20), but not enough to make it dominate the left-hand term. All counted, we still have an expected number of bits transmitted that is upper-bounded by $(3\xi^3 \lg \xi)m^2 + O(m\log m)$ whenever all the n_i's and d_i's are upper-bounded by some constant ξ, which again is on the order of m^2.

5. Discussion and Open Problems

We have introduced and studied the general problem of sampling a discrete probability distribution characterized by parameters that are scattered in remote locations. Our main goal was to minimize the amount of communication required to solve this conundrum. We considered both the unrealistic model in which arithmetic can be carried out with infinite precision and continuous random variables can be sampled exactly, and the more reasonable *random bit model* studied by Knuth and Yao [21], in which all arithmetic is carried out with finite precision and the only source of randomness comes from independent tosses of a fair coin. For a small increase in the amount of communication, we can fine-tune our technique to require only twice the number of random bits that would be provably required in the standard context in which all the parameters defining the probability distribution would be available in a single location, provided the entropy of the distribution is close to maximal.

When our framework is applied to the problem of simulating quantum entanglement with classical communication in its essentially most general form, we find that an expected number of $O(m^2)$ bits of communication suffices when there are m participants and each one of them (in the simulated world) is given an arbitrary quantum system of bounded dimension and asked to perform an arbitrary generalized measurement (POVM) with a bounded number of possible outcomes. This result generalizes and supersedes the best approach previously known in the context of multi-party entanglement, which was for the simulation of the m-partite GHZ state under projective measurements [1]. Our technique also applies without the boundedness condition on the dimension of individual systems and the number of possible outcomes per party, provided those parameters remain finite.

It would be preferable if we could eliminate the dependency of the expected number of bits of communication on the number of possible measurement outcomes. Is perfect simulation possible at all when that number is infinite, regardless of communication efficiency, a scenario in which our approach cannot be applied? In the bipartite case, Serge Massar, Dave Bacon, Nicolas Cerf, and Richard Cleve proved that classical communication can serve to simulate the effect of arbitrary measurements on maximally entangled states in a way that does not require any bounds on the number of possible outcomes [6]. More specifically, they showed that arbitrary POVMs on systems of n Bell states can be simulated with an expectation of $O(n2^n)$ bits of communication. However, their approach exploits the equivalence of this problem with a variant known as *classical teleportation* [5], in which one party has full knowledge of the quantum state and the other has full knowledge of the measurement to be applied to that state. Unfortunately, the equivalence between those two problems breaks down in a multipartite scenario and there is no obvious way to extend the approach. We leave as an open question the possibility of a simulation protocol in which the expected amount of communication would only depend on the number of participants and the dimension of their simulated quantum systems.

Our work leaves several additional important questions open. Recall that our approach provides a bounded amount on the *expected* communication required to perform exact remote sampling.

The most challenging open question is to determine if it is possible to achieve the same goal with a guaranteed bounded amount of communication *in the worst case*. If possible, this would certainly require the participants to share ahead of time the realization of random variables, possibly even continuous ones. Furthermore, a radically different approach would be needed since we had based ours on the von Neumann rejection algorithm, which has intrinsically no worst-case upper bound on its performance. This task may seem hopeless, but it has been shown to be possible for special cases of entanglement simulation in which the remote parameters are taken from a continuum of possibilities [3,8], despite earlier "proofs" that it is impossible [2].

A much easier task would be to consider other communication models, in which communication is no longer restricted to being between a single leader and various custodians. Would there be an advantage in communicating through the edges of a complete graph? Obviously, the maximum possible savings in terms of communication would be a factor of 2 since any time one participant wants to send a bit to some other participant, he can do so via the leader. However, if we care not only about the total number of bits communicated, but also the *time* it takes to complete the protocol in a realistic model in which each party is limited to sending and receiving a fixed number of bits at any given time step, parallelizing communication could become valuable. We had already shown in Ref. [1] that a parallel model of communication can dramatically improve the time needed to sample the m-partite GHZ distribution. Can this approach be generalized to arbitrary remote sampling settings?

Finally, we would like to see applications for remote sampling outside the realm of quantum information science.

Author Contributions: According to the tradition in our field, the authors are listed in alphabetical order. Conceptualization, G.B., L.D. and C.G.; Formal Analysis, G.B. and C.G.; Supervision, G.B. and L.D.; Validation, L.D.; Writing—Original Draft, C.G.; Writing—Review and Editing, G.B. and C.G.

Funding: The work of G.B. is supported in part by the Canadian Institute for Advanced Research, the Canada Research Chair program, Canada's Natural Sciences and Engineering Research Council (NSERC) and Québec's Institut transdisciplinaire d'information quantique. The work of L.D. is supported in part by NSERC.

Acknowledgments: The authors are very grateful to Nicolas Gisin for his interest in this work and the many discussions we have had with him on this topic in the past decade. Marc Kaplan provided important insights in earlier joint work on the simulation of entanglement. We also acknowledge useful suggestions provided by the anonymous referees, including the suggestion to look into Ref. [6].

Conflicts of Interest: The authors declare no conflict of interest. The funding sponsors had no role in the design of the study, in the writing of the manuscript, and in the decision to publish the results.

Abbreviations

The following abbreviations are used in this manuscript:

i.i.d. independent identically distributed
GHZ   Greenberger–Horne–Zeilinger
POVM positive-operator valued measure

References

1. Brassard, G.; Devroye, L.; Gravel, C. Exact classical simulation of the quantum-mechanical GHZ distribution. *IEEE Trans. Inf. Theory* **2016**, *62*, 876–890. [CrossRef]
2. Maudlin, T. Bell's inequality, information transmission, and prism models. In *PSA: Proceedings of the Biennial Meeting of the Philosophy of Science Association*; The University of Chicago Press: Chicago, IL, USA, 1992; pp. 404–417.
3. Brassard, G.; Cleve, R.; Tapp, A. Cost of exactly simulating quantum entanglement with classical communication. *Phys. Rev. Lett.* **1999**, *83*, 1874–1877. [CrossRef]
4. Steiner, M. Towards quantifying non-local information transfer: Finite-bit non-locality. *Phys. Lett. A* **2000**, *270*, 239–244. [CrossRef]
5. Cerf, N.J.; Gisin, N.; Massar, S. Classical teleportation of a quantum bit. *Phys. Rev. Lett.* **2000**, *84*, 2521–2524, doi:10.1103/PhysRevLett.84.2521. [CrossRef] [PubMed]

6. Massar, S.; Bacon, D.; Cerf, N.J.; Cleve, R. Classical simulation of quantum entanglement without local hidden variables. *Phys. Rev. A* **2001**, *63*, 052305, doi:10.1103/PhysRevA.63.052305. [CrossRef]
7. Gisin, N.; Gisin, B. A local variable model for entanglement swapping exploiting the detection loophole. *Phys. Lett. A* **2002**, *297*, 279–284, doi:10.1016/S0375-9601(02)00428-0. [CrossRef]
8. Toner, B.; Bacon, D. Communication cost of simulating Bell correlations. *Phys. Rev. Lett.* **2003**, *91*, 187904. [CrossRef] [PubMed]
9. Pironio, S. Violations of Bell inequalities as lower bounds on the communication cost of nonlocal correlations. *Phys. Rev. A* **2003**, *68*, 062102, doi:10.1103/PhysRevA.68.062102. [CrossRef]
10. Degorre, J.; Laplante, S.; Roland, J. Simulating quantum correlations as a distributed sampling problem. *Phys. Rev. A* **2005**, *72*, 062314, doi:10.1103/PhysRevA.72.062314. [CrossRef]
11. Shi, Y.; Zhu, Y. Tensor norms and the classical communication complexity of nonlocal quantum measurement. *SIAM J. Comput.* **2008**, *38*, 753–766, doi:10.1137/050644768. [CrossRef]
12. Degorre, J.; Kaplan, M.; Laplante, S.; Roland, J. The communication complexity of non-signaling distributions. In *Proceedings of Mathematical Foundations of Computer Science*; Královič, R., Niwiński, D., Eds.; Springer: Berlin/Heidelberg, Germany, 2009; pp. 270–281.
13. Regev, O.; Toner, B. Simulating quantum correlations with finite communication. *SIAM J. Comput.* **2009**, *39*, 1562–1580. [CrossRef]
14. Vértesi, T.; Bene, E. Lower bound on the communication cost of simulating bipartite quantum correlations. *Phys. Rev. A* **2009**, *80*, 062316, doi:10.1103/PhysRevA.80.062316. [CrossRef]
15. Bancal, J.-D.; Branciard, C.; Gisin, N. Simulation of equatorial von Neumann measurements on GHZ states using nonlocal resources. *Adv. Math. Phys.* **2010**, *2010*, 293245. [CrossRef]
16. Branciard, C.; Gisin, N. Quantifying the nonlocality of Greenberger-Horne-Zeilinger quantum correlations by a bounded communication simulation protocol. *Phys. Rev. Lett.* **2011**, *107*, 020401. [CrossRef] [PubMed]
17. Brassard, G.; Kaplan, M. Simulating equatorial measurements on GHZ states with finite expected communication cost. In Proceedings of the 7th Conference on Theory of Quantum Computation, Communication, and Cryptography (TQC), Tokyo, Japan, 17–19 May 2012; pp. 65–73.
18. von Neumann, J. Various techniques used in connection with random digits. *Monte Carlo Methods. Natl. Bur. Stand.* **1951**, *12*, 36–38.
19. Greenberger, D.M.; Horne, M.A.; Zeilinger, A. Going beyond Bell's theorem. In *Bell's Theorem, Quantum Theory and Conceptions of the Universe*; Kafatos, M., Ed.; Kluwer Academic: Dordrecht, The Netherlands, 1989; pp. 69–72.
20. Devroye, L. *Non-Uniform Random Variate Generation*; Springer: New York, NY, USA, 1986.
21. Knuth, D.E.; Yao, A.C.-C. The complexity of nonuniform random number generation. In *Algorithms and Complexity: New Directions and Recent Results*; Traub, J.F., Ed.; Academic Press: New York, NY, USA, 1976; pp. 357–428.
22. Holevo, A.S. *Statistical Structure of Quantum Theory*; Springer: New York, NY, USA, 2001.
23. Bell, J.S. On the Einstein-Podolsky-Rosen paradox. *Physics* **1964**, *1*, 195–200. [CrossRef]
24. Horn, R.A.; Johnson, C.R. *Matrix Analysis*; Cambridge University Press: Cambridge, UK, 2012.

© 2019 by the authors. Licensee MDPI, Basel, Switzerland. This article is an open access article distributed under the terms and conditions of the Creative Commons Attribution (CC BY) license (http://creativecommons.org/licenses/by/4.0/).

Article
A Classical Interpretation of the Scrooge Distribution

William K. Wootters

Department of Physics, Williams College, Williamstown, MA 01267, USA; william.wootters@williams.edu
Tel.: +1-413-597-2156

Received: 25 June 2018; Accepted: 15 August 2018; Published: 20 August 2018

Abstract: The Scrooge distribution is a probability distribution over the set of pure states of a quantum system. Specifically, it is the distribution that, upon measurement, gives up the *least* information about the identity of the pure state compared with all other distributions that have the same density matrix. The Scrooge distribution has normally been regarded as a purely quantum mechanical concept with no natural classical interpretation. In this paper, we offer a classical interpretation of the Scrooge distribution viewed as a probability distribution over the probability simplex. We begin by considering a real-amplitude version of the Scrooge distribution for which we find that there is a non-trivial but natural classical interpretation. The transition to the complex-amplitude case requires a step that is not particularly natural but that may shed light on the relation between quantum mechanics and classical probability theory.

Keywords: subentropy; GAP measure; accessible information

1. Introduction

In the early days of quantum information theory, the term "quantum communication" would typically have been understood to refer to the transmission of *classical* information via quantum mechanical signals. Such communication can be done in a sophisticated way, with the receiver making joint measurements on several successive signal particles [1,2], or it can be done in a relatively straightforward way with the receiver performing a separate measurement on each individual signal particle. In both cases, but especially in the latter case, a particularly interesting quantity, given an ensemble of quantum states to be used as an alphabet, is the ensemble's *accessible information*. This is the maximum amount of information that one can obtain about the identity of the state, on average, by making a measurement on the system described by the specified ensemble. The average here is over the outcomes of the measurement, and the maximization is over all possible measurements. In general, accessible information can be defined for ensembles consisting of pure and mixed states, but in this paper, we consider only pure-state ensembles.

Any ensemble $\{(|\psi_j\rangle, p_j)\}$ of pure quantum states with their probabilities has a unique density matrix. However, for any given density matrix ρ representing more than a single pure state, there are infinitely many ensembles—"ρ-ensembles"—described by that density matrix. Thus, it is natural to ask the following question: for a given density matrix ρ, what pure-state ρ-ensemble has the greatest value of the accessible information and what pure-state ρ-ensemble has the lowest value? The former question was answered by an early (1973) result in quantum information theory [3]—the pure-state ρ-ensemble with the greatest accessible information is the one consisting of the eigenstates of ρ with weights given by the eigenvalues. The latter question was answered in a 1994 paper [4], in which the ρ-ensemble minimizing the accessible information was called the Scrooge ensemble, or Scrooge distribution, since it is the ensemble that is most stingy with its information.

To see a simple example, consider a spin-1/2 particle whose density matrix ρ has the $|\!\uparrow\rangle$ and $|\!\downarrow\rangle$ states as its eigenvectors, with eigenvalues $\lambda_\uparrow$ and $\lambda_\downarrow$. The eigenstate ensemble for ρ, that is, the ρ-ensemble from which one can extract the most information, is the two-state ensemble consisting of the $|\uparrow\rangle$ state with probability $\lambda_\uparrow$ and the $|\downarrow\rangle$ state with probability $\lambda_\downarrow$. The optimal measurement in this case—the measurement that provides the most information—is the up-down measurement, and the amount of information it provides is equal to the von Neumann entropy of the density matrix:

$$I = S(\rho) = -(\lambda_\uparrow \ln \lambda_\uparrow + \lambda_\downarrow \ln \lambda_\downarrow). \tag{1}$$

On the other hand, the Scrooge ensemble for this density matrix is represented by a continuous probability distribution over the whole surface of the Bloch sphere. If $\lambda_\uparrow$ is larger than $\lambda_\downarrow$, then this continuous distribution is weighted more heavily towards the top of the sphere. We can write the Scrooge distribution explicitly in terms of the variable $x = (1 + \cos\theta)/2$, where θ is the angle measured from the north pole:

$$\sigma(x) = \frac{2}{\lambda_\uparrow \lambda_\downarrow} \cdot \frac{1}{\left(\frac{x}{\lambda_\uparrow} + \frac{1-x}{\lambda_\downarrow}\right)^3}. \tag{2}$$

The probability density $\sigma(x)$ is normalized in the sense that $\int_0^1 \sigma(x)dx = 1$ (the distribution is uniform over the azimuthal angle). Again, this is the ensemble of pure states from which one can extract the least information about the identity of the pure state, among all ensembles with the density matrix ρ. Somewhat remarkably, the average amount of information one gains by measuring this particular ensemble is entirely independent of the choice of measurement, as long as the measurement is complete—that is, as long as each outcome is associated with a definite pure state. This amount of information comes out to be a quantity called the subentropy Q of the density matrix:

$$I = Q(\rho) = -\frac{\lambda_\uparrow^2 \ln \lambda_\uparrow - \lambda_\downarrow^2 \ln \lambda_\downarrow}{\lambda_\uparrow - \lambda_\downarrow}. \tag{3}$$

We give more general expressions for both the Scrooge ensemble and the subentropy in Section 2 below.

In recent years, the Scrooge distribution has made other appearances in the physics literature. Of particular interest is the fact that this distribution has emerged from an entirely different line of investigation, in which the system under consideration is entangled with a large environment and the whole system is in a pure state. In that case, if one looks at the *conditional* pure states of the original system relative to the elements of an orthogonal basis of the environment, one typically finds that these conditional states are distributed by a Scrooge distribution [5–8]. In this context, the distribution is usually called a GAP measure (Gaussian adjusted projected measure, the three adjectives corresponding to the three steps by which the measure can be constructed). On another front, the Scrooge distribution has been used to address the difficult problem of bounding the *locally* accessible information when there is more than one receiver [9].

Meanwhile, the concept of subentropy, which originally arose (though without a name) in connection with the outcome entropy of random measurements [10,11], has appeared not only in problems concerning the acquisition of classical information [12–14], but also in the quantification of entanglement [15] and the study of quantum coherence [16–19]. Many detailed properties of subentropy have now been worked out, especially concerning its relation to the Shannon entropy [20–24].

Though it is possible to devise a strictly classical situation in which subentropy arises [22], the Scrooge distribution has generally been regarded as a purely quantum mechanical concept. It is, after all, a probability distribution over pure quantum states. The aim of this paper is to provide a classical interpretation of the Scrooge distribution, and in this way, to provide a new window into the relation between quantum mechanics and classical probability theory.

We find that it is much easier to make the connection if we begin by considering not the standard Scrooge distribution, but rather the analogous distribution one obtains for the case of quantum theory with *real* amplitudes. In that case, the dimension of the set of pure states is the same as the dimension of the associated probability simplex, and we find that there is a fairly natural distribution within classical probability theory that is essentially identical to the real-amplitude version of the Scrooge distribution. This distribution arises as the solution to a certain classical communication problem that we describe in Section 4.

With this interpretation of the real-amplitude Scrooge distribution in hand, we ask how the classical communication scenario might be modified to arrive at the original Scrooge ensemble for standard, complex-amplitude quantum theory. As we will see, the necessary modification is not particularly natural, but it is simple.

Thus, we begin in Sections 2 and 3 by reviewing the derivation of the Scrooge distribution and by working out the analogous distribution for the case of real amplitudes. Then, in Section 4, we set up and analyze the classical communication problem that, as we show in Section 5, gives rise to a distribution that is equivalent to the real-amplitude Scrooge distribution. In Section 6, we modify the classical communication scenario to produce the standard, complex-amplitude Scrooge distribution. Finally, we summarize and discuss our results in Section 7.

2. The Scrooge Distribution

There are several ways in which one can generate the Scrooge distribution. In this section, we review the main steps of the derivation given in Ref. [4], which applies to a Hilbert space of finite dimension. (The distribution can also be defined for an infinite-dimensional Hilbert space [5–8].) We begin by setting up the problem.

We imagine the following scenario. One participant, Alice, prepares a quantum system with an n-dimensional Hilbert space in a pure state $|x\rangle$ and sends it to Bob. Bob then tries to gain information about the identity of this pure state. Initially, Bob's state of knowledge is represented by a probability density $\sigma(x)$ over the set of pure states. (The symbol x represents a multi-dimensional parameterization of the set of pure states.) Bob makes a measurement on the system and thereby gains information. The amount of information he gains may depend on the outcome he obtains, so we are interested in the *average* amount of information he gains about x, the average being over all outcomes.

The standard quantification of Bob's average gain in information is the Shannon mutual information between the identity of the pure state and the outcome of the measurement. We can express this mutual information in terms of two probability functions: (i) the probability $p(j|x)$ of the outcome j when the state is $|x\rangle$, and (ii) the overall probability $p(j) = \int \sigma(x) p(j|x) dx$ of the outcome j averaged over the whole ensemble. In terms of these functions, the mutual information is

$$I = -\sum_j p(j) \ln p(j) + \int \sigma(x) \left[\sum_j p(j|x) \ln p(j|x) \right] dx. \tag{4}$$

The *accessible information* of the ensemble defined by $\sigma(x)$ is the maximum value of the mutual information I, where the maximum is taken over all possible measurements.

Again, for a given density matrix ρ, the Scrooge distribution is defined to be the pure-state ρ-ensemble with the lowest value of the accessible information. One can obtain the Scrooge distribution via the following algorithm [4].

We start by recalling the concept of "ρ distortion." Consider for now a finite ensemble $\{(|\psi_i\rangle, p_i)\}$ of pure states ($i = 1, \ldots, m$) whose density matrix is the completely mixed state:

$$\sum_{i=1}^{m} p_i |\psi_i\rangle \langle \psi_i| = \frac{1}{n} I. \tag{5}$$

Let $|\tilde{\psi}_i\rangle$ be the subnormalized state vector $|\tilde{\psi}_i\rangle = \sqrt{p_i} |\psi_i\rangle$, so that

$$\sum_{i=1}^{m} |\tilde{\psi}_i\rangle \langle \tilde{\psi}_i| = \frac{1}{n} I. \tag{6}$$

Under ρ distortion, each vector $|\tilde{\psi}\rangle$ is mapped to another subnormalized vector $|\tilde{\phi}\rangle$ defined by

$$|\tilde{\phi}_i\rangle = \sqrt{n\rho}|\tilde{\psi}_i\rangle. \tag{7}$$

Note that the density matrix formed by the $|\tilde{\phi}_i\rangle$'s is ρ:

$$\sum_{i=1}^{m} |\tilde{\phi}_i\rangle\langle\tilde{\phi}_i| = \sqrt{n\rho}\left(\frac{1}{n}I\right)\sqrt{n\rho} = \rho. \tag{8}$$

In terms of normalized vectors, the new ensemble is $\{(|\phi_i\rangle, q_i)\}$, with the new probabilities q_i equal to

$$q_i = \langle\tilde{\phi}_i|\tilde{\phi}_i\rangle = np_i\langle\psi_i|\rho|\psi_i\rangle. \tag{9}$$

In this way, any ensemble having the completely mixed density matrix can be mapped to a "ρ distorted" ensemble with a density matrix ρ.

The Scrooge ensemble is a continuous ensemble, not a discrete one, but the concept of ρ distortion can be immediately extended to the continuous case, and the Scrooge distribution can be easily characterized in those terms; it is the ρ distortion of the *uniform* distribution over the unit sphere in Hilbert space. The uniform distribution is the unique probability distribution over the set of pure states that is invariant under all unitary transformations.

Let us see how the ρ distortion works out in this case. First, for the uniform distribution, it is convenient to label the parameters of the pure states by y instead of x, so that we can reserve x for the Scrooge distribution. Let $\tau(y)$ be the probability density over y that represents the uniform distribution over the unit sphere (a particular parameterization will be specified shortly). In terms of normalized states, a ρ distortion maps each pure state $|y\rangle$ into the pure state $|x\rangle$ defined by

$$|x\rangle = \frac{\sqrt{\rho}|y\rangle}{\sqrt{\langle y|\rho|y\rangle}}. \tag{10}$$

This mapping defines x as a function of y: $x = f(y)$. (We write f explicitly below.) The resulting probability density over x is obtained from the continuous version of Equation (9).

$$\sigma(x) = n\tau(y)\langle y|\rho|y\rangle \mathcal{J}(y/x). \tag{11}$$

Here, $\mathcal{J}(y/x)$ is the Jacobian of the y variables with respect to the x variables. On the right-hand side of Equation (11), each y is interpreted as $f^{-1}(x)$, so that we get an expression that depends only on x.

To get an explicit expression for the Scrooge distribution—that is, an explicit expression for the probability density $\sigma(x)$—we need to choose a specific set of parameters labeling the pure states. We choose the same set of parameters to label both the uniform distribution (where we call the parameters y) and the Scrooge distribution (where we call the parameters x). We define our parameters relative to a set of normalized eigenstates $|e_j\rangle$ of the density matrix ρ. A general pure state $|x\rangle$ can be written as

$$|x\rangle = \sum_{j=1}^{n} a_j e^{-i\theta_j}|e_j\rangle, \tag{12}$$

where each a_j is a non-negative real number, and each phase θ_j runs from zero to 2π. For definiteness, employing the freedom to choose an overall phase, we define θ_n to be zero. We take x (or y) to consist of the following parameters: the squared amplitudes $x_j = a_j^2$ for $j = 1, \ldots, n-1$, and the phases θ_j for $j = 1, \ldots, n-1$. This set of $2n - 2$ parameters uniquely identifies any pure state. Later, we also use the symbol $x_n = 1 - x_1 - \cdots - x_{n-1}$. Note that the x_js are the probabilities of the outcomes of a particular orthogonal measurement associated with the eigenstates of ρ.

In terms of these parameters, the uniform distribution over the unit sphere takes a particularly simple form: it is the product of a uniform distribution over the phases and a uniform distribution over the $(n-1)$-dimensional probability simplex whose points are labeled by $\{x_1, \ldots, x_{n-1}\}$ [25]. The Scrooge distribution will likewise be a product and will be uniform over the phases but will typically have a certain bias over the probability simplex. Because the phases are always independent and uniformly distributed in the cases we consider, we omit the phases in our distribution expressions, writing the probability densities as functions of $\{x_1, \ldots, x_{n-1}\}$ (or $\{y_1, \ldots, y_{n-1}\}$).

Our aim now is to find explicit expressions for each of the factors appearing on the right-hand side of Equation (11). Since the uniform distribution over the unit sphere induces a uniform distribution over the probability simplex, the corresponding probability density $\tau(y)$ is a constant function, with the value of the constant being $(n-1)!$ as required by normalization:

$$(n-1)! \int_0^1 \int_0^{1-y_1} \cdots \int_0^{1-y_1-\cdots-y_{n-2}} dy_{n-1} \cdots dy_2 dy_1 = 1. \tag{13}$$

The function $f(y)$ defined by the ρ-distortion map, Equation (10), is given by

$$x_j = \frac{\lambda_j y_j}{\lambda_1 y_1 + \cdots + \lambda_n y_n}, \quad j = 1, \ldots, n-1, \tag{14}$$

where the λ_j's are the eigenvalues of the density matrix ρ. One finds that the inverse map is

$$y_j = \frac{x_j/\lambda_j}{x_1/\lambda_1 + \cdots + x_n/\lambda_n}, \tag{15}$$

and the Jacobian is

$$\mathcal{J}(y/x) = \frac{1}{\lambda_1 \cdots \lambda_n} \cdot \frac{1}{\left(\frac{x_1}{\lambda_1} + \cdots + \frac{x_n}{\lambda_n}\right)^n}. \tag{16}$$

Meanwhile, the factor $\langle y|\rho|y\rangle$ can be written as

$$\langle y|\rho|y\rangle = \lambda_1 y_1 + \cdots + \lambda_n y_n = \frac{1}{\frac{x_1}{\lambda_1} + \cdots + \frac{x_n}{\lambda_n}}. \tag{17}$$

By substituting the expressions from Equations (16) and (17) into Equation (11), we finally arrive at the probability density defining the Scrooge distribution:

$$\sigma(x) = \frac{n!}{\lambda_1 \cdots \lambda_n} \cdot \frac{1}{\left(\frac{x_1}{\lambda_1} + \cdots + \frac{x_n}{\lambda_n}\right)^{n+1}}. \tag{18}$$

This probability density is normalized in the sense that the integral over the probability simplex is unity:

$$\int_0^1 \int_0^{1-x_1} \cdots \int_0^{1-x_1-\cdots-x_{n-2}} \sigma(x) dx_{n-1} \cdots dx_2 dx_1 = 1. \tag{19}$$

Now, how do we know that the distribution given by Equation (18) minimizes the amount of accessible information? First, one can show that for this distribution the mutual information I is independent of the choice of measurement as long as the measurement is complete [4]. So, one can compute the value of the accessible information by considering any such measurement, and the easiest one to consider is the orthogonal measurement along the eigenstates. The result is

$$\text{accessible information} = -\sum_{k=1}^n \left(\prod_{l \neq k} \frac{\lambda_k}{\lambda_k - \lambda_l} \right) \lambda_k \ln \lambda_k, \tag{20}$$

which defines the subentropy Q. One can also show that for *any* ρ-ensemble, the *average* mutual information over all complete orthogonal measurements is equal to Q, which implies that Q is always a lower bound on the accessible information. Since the Scrooge distribution *achieves* the value Q, it achieves the minimum possible accessible information among all ρ-ensembles.

3. The Real-Amplitude Analog of the Scrooge Distribution

Though our own world is described by standard quantum theory with complex amplitudes, we can also consider an analogous, hypothetical theory with real amplitudes. A pure state in the real-amplitude theory is represented by a real unit vector, and a density matrix is represented by a symmetric real matrix with non-negative eigenvalues and unit trace. Time evolution in this theory is generated by an antisymmetric real operator in place of the antihermitian operator iH.

The question considered in the preceding section can also be asked in regard to the real-amplitude theory. Given a density matrix ρ, we ask what ρ-ensemble has the smallest value of accessible information. It turns out that essentially all of the methods used in the preceding section continue to work in the real case. Again one begins with the uniform distribution over the unit sphere of pure states, and again, one obtains the Scrooge ensemble (in this case the real-amplitude Scrooge ensemble) via ρ distortion. The arguments leading to the conclusion that the ensemble produced in this way minimizes the accessible information work just as well in the real-amplitude case as in the complex-amplitude case.

The one essential difference between the two cases lies in the form of the initial probability density $\tau(y)$ that is associated with the uniform distribution over the unit sphere in Hilbert space. Whereas in the complex case the induced distribution over the probability simplex is uniform, in the real case, the induced distribution over the probability simplex is more heavily weighted towards the edges and corners.

We can see an example by considering the case with $n = 2$. Instead of starting with a uniform distribution over the surface of the Bloch sphere, one starts with a uniform distribution over the unit circle in a two-dimensional real vector space. Let γ be the angle around this circle measured from some chosen axis (once a density matrix has been specified, we will take this axis to be along one of the eigenstates of the density matrix). Then, γ is initially uniformly distributed. The parameter analogous to y_1 of the preceding section is $y = \sin^2 \gamma$. Note that y runs from 0 to 1 as γ runs from 0 to $\pi/2$. The initial probability density $\tau_r(y)$ is therefore obtained from

$$\tau_r(y)dy = (2/\pi)d\gamma, \tag{21}$$

which leads to

$$\tau_r(y) = \frac{1}{\pi} \cdot \frac{1}{\sqrt{y(1-y)}} \tag{22}$$

(the subscript r represents "real"). This is in contrast to the function $\tau(y) = 1$ that would apply in the complex-amplitude case. We see that in the real case, $\tau_r(y)$ is largest around $y = 0$ and $y = 1$.

For n dimensions, we take as our parameters specifying a pure state (i) the first $n-1$ probabilities y_j ($j = 1, \ldots, n-1$) of the outcomes of a certain orthogonal measurement (which we will choose to be the measurement along the eigenvectors of the given density matrix), and (ii) a set of discrete phase parameters (each of them taking the values ± 1), which will always be independently and uniformly distributed and therefore suppressed in our expressions for the probability densities.

For the uniform distribution over the unit sphere in the n-dimensional real Hilbert space, one can show that the induced distribution over the parameters $(y_1, \ldots, y_{n-1})$ is given by [26]

$$\tau_r(y) = \frac{\Gamma(n/2)}{\pi^{n/2}} \cdot \frac{1}{\sqrt{y_1 \cdots y_n}}, \tag{23}$$

where $y_n = 1 - y_1 - \cdots - y_{n-1}$. This probability density is normalized over the probability simplex, as in Equation (19):

$$\int_0^1 \int_0^{1-y_1} \cdots \int_0^{1-y_1-\cdots-y_{n-2}} \tau_r(y) dy_{n-1} \cdots dy_2 dy_1 = 1. \tag{24}$$

The general expression for $\sigma(x)$ given in Equation (11) remains valid in the real case, as do Equations (15)–(17) for the various factors in Equation (11). Again, the one difference is in $\tau_r(y)$, for which we now use Equation (23). By combining these ingredients, we arrive at our expression for the real-amplitude Scrooge ensemble:

$$\sigma_r(x) = \frac{n\Gamma(n/2)}{\pi^{n/2}\sqrt{\lambda_1 \cdots \lambda_n}\sqrt{x_1 \cdots x_n}\left(\frac{x_1}{\lambda_1} + \cdots + \frac{x_n}{\lambda_n}\right)^{\frac{n}{2}+1}}, \tag{25}$$

where, as before, the λ_j's are the eigenvalues of the density matrix whose Scrooge distribution is being computed.

Though Equation (25) was derived as a distribution over the set of pure states in real-amplitude quantum theory, it reads as a probability distribution over the $(n-1)$-dimensional probability simplex for a classical random variable with n possible values. One can therefore at least imagine that there might be a classical scenario in which this distribution is natural. In the following section, we identify such a scenario.

4. Communicating with Dice

Ref. [26] imagined the following classical communication scenario. Alice is trying to convey to Bob the location of a point in an $(n-1)$-dimensional probability simplex. To do this, she constructs a weighted n-sided die that, for Bob, has the probabilities corresponding to the point that Alice is trying to convey. She then sends the die to Bob, who rolls the die many times in order to estimate the probabilities of the various possible outcomes. However, the information transmission is limited in that Bob is allowed only a fixed number of rolls—let us call this number N (perhaps the die automatically self-destructs after N rolls). So, Bob will always have an imperfect estimate of the probabilities that Alice is trying to convey. Alice and Bob are allowed to choose in advance a discrete set of points in the probability simplex—these are the points representing the set of signals Alice might try to send—and they choose this set of points, along with their *a priori* weights, so as to maximize the mutual information between the identity of the point being conveyed and the result of Bob's rolls of the die. The main result of that paper was that in the limit of a large N, the optimal distribution of points in the probability simplex approximates the continuous distribution over the simplex expressed by the following probability density:

$$\hat{\tau}(y) = \frac{\Gamma(n/2)}{\pi^{n/2}} \cdot \frac{1}{\sqrt{y_1 \cdots y_n}}, \tag{26}$$

where the y_js are the probabilities (we use a hat in our labels of probability densities that arise in a classical context). This result is interesting because it is the same probability density as the one induced by the uniform distribution over the unit sphere in real Hilbert space (Equation (23) above). Thus, in a world based on real-amplitude quantum theory as opposed to the complex-amplitude theory, there is a sense in which one could say that nature optimizes the transfer of information.

That paper—and closely related papers [27,28]—deal only with the uniform distribution over the unit sphere, not with non-trivial Scrooge distributions. In the present section, we consider a modification of the above communication scenario, and in the next section, we show that this modified scheme yields the real-amplitude Scrooge distribution.

A natural way to generalize the above communication scheme is this: let the allowed number N of rolls of the die vary from one die to another (that is, some dice last longer than others before they self-destruct). Now, once N is allowed to vary, it makes sense to let N itself be another random variable that conveys information. We are thus led to consider the following scenario.

Alice is trying to convey to Bob an ordered n-tuple of non-negative real numbers $(M_1, \ldots, M_n)$ (Alice and Bob agree in advance on a specific set of such ordered n-tuples, any one of which Alice might try to convey). Let us refer to such an n-tuple as a "signal." In order to convey her signal, Alice sends Bob an n-sided die that Bob then begins to roll over and over, keeping track of the number of times each outcome occurs. N_j is the number of times that the outcome j occurs. At some point, the die self-destructs. Alice has constructed both the weighting of the die and the self-destruction mechanism so that the *average* value of N_j is M_j.

However, both the rolling of the die and its duration are probabilistic, and Alice cannot completely control either the individual numbers N_j or their sum. For any given signal $(M_1, \ldots, M_n)$, we assume that each N_j is distributed independently according to a Poisson distribution with mean value M_j:

$$P(N_1, \ldots, N_n | M_1, \ldots, M_n) = \prod_{j=1}^{n} e^{-M_j} \frac{M_j^{N_j}}{N_j!}. \tag{27}$$

This is equivalent to assuming that the *total* number N of rolls of the die is Poisson distributed with a mean value of $M = M_1 + \cdots + M_n$ and that for a given total number of rolls, the numbers of occurrences of the individual outcomes are distributed according to a multinomial distribution with weights M_j/M. That is, we are assuming the usual statistics for rolling a die, together with a Poisson distribution for the total number of rolls (another model we could have used is to have Alice send Bob a radioactive sample that can decay in n ways and that Bob is allowed to observe with detectors for a fixed amount of time).

To make the problem interesting, and to keep Alice from being able to send Bob an arbitrarily large amount of information in a single die, limits are placed on the sizes of $M_1, \ldots, M_n$. This is done by imposing, for each j, an upper bound $\mathcal{M}_j$ (script M) on the expectation value of the number of times the j outcome occurs. This expectation value is an average over all the possible signals that Alice might send.

We also need to say in what sense Alice and Bob are optimizing their communication. There are a number of reasonable options for doing this—e.g., we could say they maximize the mutual information, or minimize the probability of error for a fixed number of signals—but it is likely that many of these formulations will be essentially equivalent when the values of $\mathcal{M}_j$ become very large. Here, we take a simple, informal approach. We say that, in order to make the various signals distinguishable from each other, Alice and Bob choose their n-tuples $(M_1, \ldots, M_n)$ so that neighboring signals, say $(M_1, \ldots, M_n)$ and $(M_1 + \Delta M_1, \ldots, M_n + \Delta M_n)$, are at least a certain distance from each other, and we use the Fisher information metric to measure distance. Specifically, we require the Fisher information distance between the probability distributions $P(N_1, \ldots, N_n | M_1, \ldots, M_n)$ and $P(N_1, \ldots, N_n | M_1 + \Delta M_1, \ldots, M_n + \Delta M_1)$ to be greater than or equal to a specified value d_{min} (or, equivalently for small $\Delta M_j / M_j$, we require the Kullback–Leibler divergence to be at least $(1/2) d_{min}^2$). For the Poisson distribution and for small values of the ratios $\Delta M_j / M_j$, this condition works out to be

$$\sum_{j=1}^{n} \frac{(\Delta M_j)^2}{M_j} \geq d_{min}^2. \tag{28}$$

For our purposes the exact value of d_{min} is not important. We also assume that the various signals have equal *a priori* probabilities. This is a natural choice if one wants to convey as much information as possible. Under these assumptions, Alice and Bob's aim is to maximize the number of distinct signals.

The analysis will be much simpler if we parameterize each die not by $(M_1, \ldots, M_n)$, but rather by the variables

$$\alpha_j = \sqrt{M_j}, \quad j = 1, \ldots, n. \tag{29}$$

Then, for neighboring signals we can write

$$\Delta\alpha_j = \frac{1}{2\sqrt{M_j}}\Delta M_j, \qquad (30)$$

so that the condition in Equation (28) becomes

$$\sum_{j=1}^{n}(\Delta\alpha_j)^2 \geq \frac{1}{4}d_{min}^2. \qquad (31)$$

That is, in the space parameterized by $\vec{\alpha} = (\alpha_1,\ldots,\alpha_n)$, we want the points representing Alice's signals to be evenly separated from each other. Thus Alice's signals will be roughly uniformly distributed over some region of $\vec{\alpha}$-space—she wants to pack in as many signals as possible without exceeding the bounds M_j on the expectation values of the N_js. In what follows, we approximate this discrete but roughly uniform distribution of the values of $\vec{\alpha}$ by a continuous probability distribution. The probability density is zero outside the region where Alice's possible signals lie; inside that region, it has a constant value of $1/V$, where V is the volume of the region.

The communication problem then becomes a straightforward geometry problem—within the "positive" section of $\vec{\alpha}$-space (that is, the section in which each α_j is non-negative), the aim is to find the region $\mathcal{R}$ of largest volume that satisfies the constraints

$$\frac{1}{V_\mathcal{R}}\int_\mathcal{R}\alpha_j^2\,d\vec{\alpha} = M_j, \quad j=1,\ldots,n, \qquad (32)$$

where $V_\mathcal{R}$ is the volume of $\mathcal{R}$. We maximize the volume because Alice's signals have a fixed packing density within $\mathcal{R}$; thus the larger the volume, the more signals Alice has at her disposal.

It is not hard to see that the solution to this geometry problem is to make region $\mathcal{R}$ the positive section of a certain ellipsoid centered at the origin. To see this, the conditions (32) can be written as

$$\int_\mathcal{R}\alpha_j^2\,d\vec{\alpha} = M_j\int_\mathcal{R}d\vec{\alpha}, \quad j=1,\ldots,n. \qquad (33)$$

Now, let $\beta_j = \alpha_j/\sqrt{M_j}$. In terms of the β_js, the above conditions become

$$\int_{\mathcal{R}'}\beta_j^2\,d\vec{\beta} = \int_{\mathcal{R}'}d\vec{\beta}, \quad j=1,\ldots,n, \qquad (34)$$

where $\mathcal{R}'$ is the region of $\vec{\beta}$-space corresponding to the region $\mathcal{R}$ of $\vec{\alpha}$-space. In particular, the equation obtained by summing these n conditions must also be true:

$$\int_{\mathcal{R}'}\beta^2\,d\vec{\beta} = n\int_{\mathcal{R}'}d\vec{\beta}, \qquad (35)$$

where $\beta^2 = \beta_1^2 + \cdots + \beta_n^2$. That is, the average squared distance from the origin over region $\mathcal{R}'$ must be equal to n. The maximum volume region $\mathcal{R}'$ satisfying this one condition is the positive section of a sphere, and one can work out that the radius of the sphere must be $\sqrt{n+2}$. Moreover, that region also satisfies all of the conditions (34). So, that same region is the maximum volume region that satisfies those conditions as well. Going back to the α_j's, we see that the maximum volume region satisfying the conditions (32) is the positive section of an ellipsoid, with semi-axis lengths

$$\alpha_j^{max} = \sqrt{(n+2)M_j}. \qquad (36)$$

Thus, the strategy that Alice and Bob adopt is to choose a set of closely packed signals with some minimum separation in $\vec{\alpha}$-space that occupies the positive section of an ellipsoid centered at the origin. Again, in this paper, we treat this discrete but roughly uniform distribution of signals as if it were actually uniform. This approximation becomes more and more reasonable as the values of the $\mathcal{M}_j$s increase.

5. A Distribution over the Probability Simplex

So far, we have not made any connection between our communication problem and the real-amplitude Scrooge distribution. We do this now by seeing how the uniform distribution over the ellipsoid in $\vec{\alpha}$-space induces a certain probability distribution over the $(n-1)$-dimensional probability simplex for Alice's n-sided die. We define this probability distribution as follows.

Let us imagine many rounds of communication from Alice to Bob: she has sent him many dice for which the expected numbers of occurrences of the various outcomes, $(M_1, \ldots, M_n)$, cover a representative range of values: the corresponding values of $\vec{\alpha}$ are distributed fairly uniformly over the region $\mathcal{R}$ in $\vec{\alpha}$-space. Bob has rolled each of these dice as many times as it can be rolled. Now consider a small region of the probability simplex, say the region $\mathcal{S}(x, \Delta x)$ for which the probability of the jth outcome lies between x_j and $x_j + \Delta x_j$ for $j = 1, \ldots, n-1$. Some of the dice Alice has sent to Bob have probabilities lying in this region. The weight we want to attach to the region $\mathcal{S}(x, \Delta x)$ is, roughly speaking, the fraction of the total number of rolls that came from dice in this region. Note that for a die at location $\vec{\alpha}$, the expectation value of the number of times it will be rolled is $\alpha^2 = \alpha_1^2 + \cdots + \alpha_n^2$. So, we multiply the density of signals by the factor α^2 to get the "density of rolls." These considerations lead us to the following definition of the weight $\hat{\sigma}(x) dx_1 \cdots dx_{n-1}$ that we assign to the infinitesimal region $\mathcal{S}(x, dx)$:

$$\hat{\sigma}(x) dx_1 \cdots dx_{n-1} = \frac{\int_{\mathcal{C}(x,dx)} \alpha^2 d\vec{\alpha}}{\int_{\mathcal{R}} \alpha^2 d\vec{\alpha}}. \tag{37}$$

Here, $\mathcal{C}(x, dx)$ is the cone (within the region $\mathcal{R}$) representing dice for which the probabilities of the outcomes lie in $\mathcal{S}(x, dx)$:

$$\mathcal{C}(x, dx) = \left\{ \vec{\alpha} \in \mathcal{R} \,\middle|\, x_j \leq \frac{\alpha_j^2}{\alpha^2} \leq x_j + dx_j \right\}. \tag{38}$$

Our use of the weighting factor α^2 is reminiscent of the "adjustment" stage in the construction of the GAP measure in Refs. [5–8], and the integration over $\mathcal{C}(x, dx)$ is reminiscent of the projection stage of that same construction. We can express $\hat{\sigma}(x)$ more formally as

$$\hat{\sigma}(x) = \frac{\int_{\mathcal{R}} \left[\prod_{j=1}^{n-1} \delta\left(x_j - \frac{\alpha_j^2}{\alpha^2}\right) \right] \alpha^2 d\vec{\alpha}}{\int_{\mathcal{R}} \alpha^2 d\vec{\alpha}}, \tag{39}$$

where δ is the Dirac delta function.

It is not difficult to obtain an explicit expression for $\hat{\sigma}(x)$ starting with Equation (39). For example, in the integral appearing in the numerator of that equation, one can use the integration variables $s_1, \ldots, s_{n-1}$ and α, where $s_j = \alpha_j/\alpha$. Then, $d\vec{\alpha}$ becomes $(1/s_n)\alpha^{n-1} ds_1 \ldots ds_{n-1} d\alpha$, and the integral becomes straightforward. Here, though, we take a different path to the same answer, starting with Equation (37). This latter approach turns out to be more parallel to our derivation of the Scrooge distribution in the quantum mechanical setting.

First, note that the numerator in Equation (37) can be written as

$$\int_{\mathcal{C}(x,dx)} \alpha^2 d\vec{\alpha} = \frac{n}{n+2} \alpha_{max}^2 \cdot (\text{volume of } \mathcal{C}(x, dx)), \tag{40}$$

where α_{max} is the largest value of α over all points in $\mathcal{R}$ satisfying $\alpha_j^2/\alpha^2 = x_j$ for $j = 1, \ldots, n$. We get Equation (40) by writing $d\vec{\alpha}$ as $k\alpha^{n-1} d\alpha$, with some constant k, for the purpose of integrating over the cone. We can find the value of α_{max} by finding the point of intersection between (i) the ellipsoid that defines the boundary of $\mathcal{R}$, given by

$$\frac{\alpha_1^2}{(n+2)\mathcal{M}_1} + \cdots + \frac{\alpha_n^2}{(n+2)\mathcal{M}_n} = 1, \tag{41}$$

and (ii) the line parameterized by α and defined by the equations

$$\alpha_j = \sqrt{x_j}\alpha, \quad j = 1, \ldots, n. \tag{42}$$

The value of α at this intersection point is

$$\alpha_{max} = \sqrt{\frac{n+2}{\frac{x_1}{\mathcal{M}_1} + \cdots + \frac{x_n}{\mathcal{M}_n}}}. \tag{43}$$

We can therefore rewrite Equation (40) as

$$\int_{\mathcal{C}(x,dx)} \alpha^2 d\vec{\alpha} = \frac{n}{\frac{x_1}{\mathcal{M}_1} + \cdots + \frac{x_n}{\mathcal{M}_n}} \cdot (\text{volume of } \mathcal{C}(x,dx)). \tag{44}$$

Meanwhile, it follows from Equation (32) that the denominator in Equation (37) is

$$\int_{\mathcal{R}} \alpha^2 d\vec{\alpha} = (\mathcal{M}_1 + \cdots + \mathcal{M}_n) V_{\mathcal{R}}. \tag{45}$$

Our next step is to compare $\hat{\sigma}(x)$ to the analogous distribution $\hat{\tau}(y)$ induced by the uniform distribution of the vector $\vec{\beta}$—the same $\vec{\beta}$ as in Section 4—over its domain $\mathcal{R}'$ (recall that $\mathcal{R}'$ is the positive section of a sphere):

$$\hat{\tau}(y) dy_1 \cdots dy_{n-1} = \frac{\int_{\mathcal{C}'(y,dy)} \beta^2 d\vec{\beta}}{\int_{\mathcal{R}'} \beta^2 d\vec{\beta}}. \tag{46}$$

Here, $\mathcal{C}'(y,dy)$ is the cone in $\mathcal{R}'$ for which $y_j \leq (\beta_j/\beta)^2 \leq y_j + dy_j$. We can immediately write down an explicit expression for $\hat{\tau}(y)$. It is the same as the distribution (23) on the probability simplex induced by the uniform distribution over the unit sphere in the n-dimensional real Hilbert space—the extra radial dimension represented by β has no bearing on the distribution over the probability simplex. Thus,

$$\hat{\tau}(y) = \frac{\Gamma(n/2)}{\pi^{n/2}} \cdot \frac{1}{\sqrt{y_1 \cdots y_n}}. \tag{47}$$

The expression for $\hat{\sigma}(x)$ is determined by finding the factors by which the numerator and denominator in Equation (46) change when the sphere in $\vec{\beta}$-space is stretched into an ellipsoid in $\vec{\alpha}$-space. In this transformation (in which $\alpha_j = \beta_j\sqrt{\mathcal{M}_j}$), the relation between y (in Equation (46)) and x (in Equation (37)) is given by $y = g(x)$, where g takes the point $(\alpha_1^2/\alpha^2, \ldots, \alpha_{n-1}^2/\alpha^2)$ in the probability simplex to the point $(\beta_1^2/\beta^2, \ldots, \beta_{n-1}^2/\beta^2)$.

Essentially, any appearance of $\mathcal{M}_j$ in our expression (37) for $\hat{\sigma}(x) dx_1 \ldots dx_{n-1}$ becomes a 1 in Equation (46). Thus, according to Equation (44), when we transform from $\vec{\beta}$ to $\vec{\alpha}$, the numerator in Equation (46) is multiplied by

$$\frac{\frac{n}{\frac{x_1}{\mathcal{M}_1} + \cdots + \frac{x_n}{\mathcal{M}_n}} \cdot (\text{volume of } \mathcal{C}(x,dx))}{n \cdot (\text{volume of } \mathcal{C}'(y,dy))}, \tag{48}$$

and according to Equation (45), in this same transformation, the denominator in Equation (46) is multiplied by

$$\frac{(\mathcal{M}_1 + \cdots + \mathcal{M}_n) V_\mathcal{R}}{n V_{\mathcal{R}'}}. \tag{49}$$

For both the transitions $\mathcal{C}'(y, dy) \to \mathcal{C}(x, dx)$ and $\mathcal{R}' \to \mathcal{R}$, the volume increases by a factor of $\sqrt{\mathcal{M}_1 \cdots \mathcal{M}_n}$. So, these volume factors cancel out. By inserting the other factors from Equations (48) and (49), it is found that

$$\hat{\sigma}(x) = \hat{\tau}(y) \mathcal{J}(y/x) \frac{n}{\left(\frac{x_1}{\mathcal{M}_1} + \cdots + \frac{x_n}{\mathcal{M}_n}\right)(\mathcal{M}_1 + \cdots + \mathcal{M}_n)}, \tag{50}$$

where $\mathcal{J}(y/x)$ is the Jacobian of y with respect to x.

Let us now write y explicitly in terms of x:

$$y_j = \frac{\beta_j^2}{\beta^2} = \frac{\frac{\alpha_j^2}{\mathcal{M}_j}}{\frac{\alpha_1^2}{\mathcal{M}_1} + \cdots + \frac{\alpha_n^2}{\mathcal{M}_n}} = \frac{\frac{x_j}{\mathcal{M}_j}}{\frac{x_1}{\mathcal{M}_1} + \cdots + \frac{x_n}{\mathcal{M}_n}}. \tag{51}$$

From this, we can get the Jacobian (very much like the one in Equation (16)):

$$\mathcal{J}(y/x) = \frac{1}{\mathcal{M}_1 \cdots \mathcal{M}_n} \cdot \frac{1}{\left(\frac{x_1}{\mathcal{M}_1} + \cdots + \frac{x_n}{\mathcal{M}_n}\right)^n}. \tag{52}$$

By inserting the results of Equations (51) and (52) into Equation (50), we arrive at

$$\hat{\sigma}(x) = \frac{n \Gamma(n/2)}{\pi^{n/2} \mathcal{M} \sqrt{\mathcal{M}_1 \cdots \mathcal{M}_n} \sqrt{x_1 \cdots x_n} \left(\frac{x_1}{\mathcal{M}_1} + \cdots + \frac{x_n}{\mathcal{M}_n}\right)^{\frac{n}{2}+1}}, \tag{53}$$

where $\mathcal{M} = \mathcal{M}_1 + \cdots + \mathcal{M}_n$. This is essentially the same as the expression (25) obtained earlier as the real-amplitude Scrooge distribution. The agreement can be made more explicit by defining the ratios $\lambda_j = \mathcal{M}_j / \mathcal{M}$, in which case Equation (53) becomes exactly identical to Equation (25), with these λ_js playing the role of the eigenvalues of the density matrix.

Note that in the above derivation, we see an analog of ρ distortion. The stretching of the sphere in $\vec{\beta}$-space into an ellipsoid in $\vec{\alpha}$-space is very much like ρ distortion, though in place of the notion of a density matrix, we have a uniform distribution within the sphere or ellipsoid.

It may seem that our communication set-up, in which Alice sends a die equipped with a probabilistic self-destruction mechanism, is rather artificial. However, the mathematics is actually fairly simple and natural. We are considering a set of Poisson-distributed random variables and are basically constructing a measure on the set of values of these variables based on distinguishability (this is the measure derived from the Fisher information metric). That measure then induces a measure on the probability simplex which agrees with the real-amplitude Scrooge distribution.

6. A Classical Interpretation of the Complex-Amplitude Scrooge Distribution

We now show how to modify the above classical communication scenario to arrive at the original, complex-amplitude Scrooge distribution.

Not surprisingly, we begin by doubling the number of sides of Alice's dice. Let the outcomes be labeled $1_a, 1_b, 2_a, 2_b, \ldots, n_a, n_b$. The communication scheme is exactly as it was in Section 4, except that instead of placing an upper bound on the expectation value of the number of times each individual outcome occurs, the j_a and j_b outcomes are grouped together and an upper bound $\mathcal{M}_j$ is placed on the expectation value of the total number of times the two j outcomes occur. This is done for each $j = 1, \ldots, n$. Again, Alice and Bob are asked to maximize the number of distinguishable signals under this constraint, where "distinguishable" again means having a Fisher-distance separation of at least d_{min}.

As before, it is easiest to view the problem in $\vec{\alpha}$-space; let us label the variables in the space α_{ja} and α_{jb}. We now look for the maximum-volume region $\mathcal{R}$ of the positive section of $\vec{\alpha}$-space satisfying the constraints

$$\frac{1}{V_\mathcal{R}} \int_\mathcal{R} (\alpha_{ja}^2 + \alpha_{jb}^2) d\vec{\alpha} = \mathcal{M}_j, \quad j=1,\ldots,n. \tag{54}$$

In terms of the variables $\beta_{ja} = \alpha_{ja}/\sqrt{\mathcal{M}_j}$ and $\beta_{jb} = \alpha_{jb}/\sqrt{\mathcal{M}_j}$, the constraints become

$$\frac{1}{V_{\mathcal{R}'}} \int_{\mathcal{R}'} (\beta_{ja}^2 + \beta_{jb}^2) d\vec{\beta} = 1, \quad j=1,\ldots,n, \tag{55}$$

where $\mathcal{R}'$ is the region in $\vec{\beta}$-space corresponding to $\mathcal{R}$. Upon summing these n constraints, the equation

$$\frac{1}{V_{\mathcal{R}'}} \int_{\mathcal{R}'} \beta^2 d\vec{\beta} = n \tag{56}$$

is obtained, where $\beta^2 = \sum_{j=1}^n (\beta_{ja}^2 + \beta_{jb}^2)$. Maximizing the volume under this constraint again gives a sphere in $\vec{\beta}$-space, which becomes an ellipsoid in $\vec{\alpha}$-space (restricted to the positive section).

Continuing as before, one finds that the induced probability distribution over the $(2n-1)$-dimensional probability simplex associated with a $2n$-sided die is the analog of Equation (53), the n values $\mathcal{M}_1,\ldots,\mathcal{M}_n$ now being replaced by the $2n$ values $\mathcal{M}_1/2, \mathcal{M}_1/2, \ldots, \mathcal{M}_n/2, \mathcal{M}_n/2$.

$$\hat{\sigma}_{ab}(x) = \frac{n\Gamma(n)}{\pi^n \lambda_1 \cdots \lambda_n \sqrt{x_{1a} x_{1b} \cdots x_{na} x_{nb}} \left(\frac{x_{1a}+x_{1b}}{\lambda_1} + \cdots + \frac{x_{na}+x_{nb}}{\lambda_n} \right)^{n+1}}, \tag{57}$$

where $\lambda_j = \mathcal{M}_j / \mathcal{M}$. Here, x_{ja} and x_{jb} are the probabilities of the outcomes j_a and j_b, and x refers to the point $(x_{1a}, x_{1b}, \ldots, x_{(n-1)a}, x_{(n-1)b}, x_{na})$ in the $(2n-1)$-dimensional probability simplex (the value of x_{nb} is determined by the requirement that the probabilities sum to unity).

Finally, a distribution over the $(n-1)$-dimensional probability simplex is obtained by ignoring the difference between the outcomes j_a and j_b. We can imagine an observer who, unlike Alice and Bob, cannot see the a and b. For this "ab-blind" observer, the distribution of Equation (57) looks like the following distribution over the $(n-1)$-dimensional probability simplex:

$$\hat{\sigma}(x) = \int \prod_{j=1}^{n-1} \delta[x_j - (x_{ja} + x_{jb})] \hat{\sigma}_{ab}(x) dx_{1a} dx_{1b} \cdots dx_{na}. \tag{58}$$

Here, δ is the Dirac delta function and the integral is over the $(2n-1)$-dimensional probability simplex. The integral in Equation (58) is straightforward, and it can be found that

$$\hat{\sigma}(x) = \frac{n!}{\lambda_1 \cdots \lambda_n} \cdot \frac{1}{\left(\frac{x_1}{\lambda_1} + \cdots + \frac{x_n}{\lambda_n} \right)^{n+1}}. \tag{59}$$

This is the same as the original Scrooge distribution of Equation (18). The role of the eigenvalues of the density matrix is now played by the set of values $\lambda_j = \mathcal{M}_j/(\mathcal{M}_1 + \cdots + \mathcal{M}_n)$, where, again, $\mathcal{M}_j$ is the maximum allowed expectation value of the number of times that the outcomes j_a and j_b occur.

7. Discussion

In this paper we have shown how the real-amplitude version of the Scrooge distribution emerges naturally from a classical communication scenario in which information is transmitted via the values of several random variables N_j. Essentially, the real-amplitude Scrooge distribution, regarded as a probability distribution over the probability simplex, is derived from an underlying distribution based on distinguishability. Our analysis includes a transformation that plays something like the role of a ρ distortion: in place of a density matrix, what is distorted is a distribution over the space of potential signals.

In order to get the original complex-amplitude Scrooge distribution for dimension n, we needed to consider a case with twice as many random variables, grouped into pairs, and then we imagined an observer for whom only the *sum* of the variables within each pair was observable.

The reader will probably have noticed that the role played by the concept of *information* in our classical communication problem seems to be exactly the opposite of the role it plays in the quantum origin of the Scrooge distribution. In quantum theory, the Scrooge distribution is the distribution over pure states that, upon measurement, provides an observer with the *least* possible amount of information. In contrast, in our classical communication scenario, the Scrooge distribution emerges from a requirement that Alice convey as much information as possible to Bob. What is common to both cases is that the information-based criterion favors a distribution that is highly *spread out* over the probability simplex. In the quantum case, a distribution spread out over many non-orthogonal states tends to make it difficult for an observer to gain information about the state. In the classical case, Alice and Bob want to spread their signals as widely as possible over the space of possibilities in order to maximize the number of distinguishable signals. Thus, though the two scenarios are quite different, their extremization criteria have similar effects.

An intriguing aspect of our classical scenario is that the probability simplex is not itself taken as the domain in which the problem is formulated. Instead, the problem is formulated in terms of the number of times each outcome occurs. The distribution over the probability simplex is a secondary concept, being derived from a more fundamental distribution over the space of the numbers of occurrences of the outcomes. That is, the M_j values are more fundamental in the problem than the probabilities of the outcomes, which are defined in terms of the M_js by the equation $x_j = M_j/M$. In this specific respect, then, the effort to find a classical interpretation of the Scrooge distribution seems to lead us away from the models studied in Refs. [26,28], in which the set of frequencies of occurrence of the measurement outcomes was the only source of information considered.

It is interesting to ask whether this feature of our scenario is necessary in order to get the Scrooge distribution classically. To address this question, in Appendix A we consider another classical communication problem, in which we impose a separate restriction for each outcome as in Section 4, but now with Alice's signals consisting purely of probabilities (which are estimated by Bob through the observed frequencies of occurrence). For simplicity, we restrict our attention to the most basic case, in which there are only two possible outcomes—so Alice's die is now a coin to be tossed—and we are aiming just for the real-amplitude Scrooge distribution as opposed to the complex-amplitude version. We find that the resulting probability distribution over the probability simplex is *not* of the same form as the real-amplitude Scrooge distribution. This result can be taken as one bit of evidence that it is indeed necessary to go beyond the probability simplex and to work in a space of one additional dimension in order to obtain the Scrooge distribution classically. In this connection, it is worth noting that something very similar has been seen in research on *subentropy*—certain simple relations between subentropy and the Shannon entropy can be obtained only by lifting the normalization restriction that defines the probability simplex and working in the larger space of unnormalized n-tuples [21,23].

Finally, one might wonder about the potential significance of our need to invoke an "*ab*-blind" observer in order to obtain the complex-amplitude Scrooge distribution. It is well known that the number of independent parameters required to specify a pure quantum state (of a system with a finite-dimensional Hilbert space) is exactly twice the number of independent probabilities associated with a complete orthogonal measurement on the system. Here, we are seeing another manifestation of this factor of two: the classical measurement outcomes, corresponding to the sides of a rolled die, have to be grouped into pairs, and we need to imagine an observer incapable of distinguishing between the elements of any pair. In our actual quantum world, one can reasonably ask whether there is any interesting sense in which we ourselves are "ab-blind." This question, though, lies well beyond the scope of the present paper.

Funding: This research received no external funding.

Conflicts of Interest: The author declares no conflict of interest.

Appendix A. Communicating through Probabilities

Here, we consider a classical communication problem based directly on probabilities, as opposed to being based on the number of times each outcome occurs. We restrict our attention to the case of two outcomes, which we imagine as "heads" and "tails" for a tossed coin. The question is whether the real-amplitude Scrooge distribution for $n = 2$ can be obtained in this way.

Alice is trying to convey to Bob the identity of a point in the one-dimensional probability simplex (not the two-dimensional space with axes labeled "number of heads" and "number of tails"). The "simplex" in this case is just a line segment, and the points of the simplex are labeled by the probability x of heads occurring (the probability of tails occurring is $1 - x$). Alice conveys her signal by sending Bob a coin with weights $(x, 1 - x)$. Bob tosses the coin in order to estimate the value of x, but he is allowed to toss it only N times, at which point the coin will self-destruct. Alice chooses a set of points in the probability simplex in advance that will serve as her potential signals, and she provides Bob with the list of these points. Alice also chooses a function $N(x)$ that determines how many times Bob will be able to toss the coin if the coin's weights are $(x, 1 - x)$. However, Bob does not know the function $N(x)$ and is not allowed to use the observed total number of tosses in his estimation of the value of x. He can use only the frequencies of occurrence of heads and tails.

We limit the amount of information that Alice can convey per coin by specifying the values of two quantities: (i) the expectation value $\mathcal{N}$ of the total number of tosses, and (ii) the expectation value $\mathcal{N}_H$ of the number of heads. If we let $\rho(x)dx$ be the number of signals lying between the values x and $x + dx$, we can write these two restrictions as follows:

$$\int_0^1 N(x)\rho(x)dx = \mathcal{N} \int_0^1 \rho(x)dx. \tag{A1}$$

$$\int_0^1 xN(x)\rho(x)dx = \mathcal{N}_H \int_0^1 \rho(x)dx. \tag{A2}$$

As before, we insist that Alice choose the signal values so that neighboring signals have a certain minimum degree of distinguishability as quantified by the Fisher information metric. For the binomial distributions we are dealing with here, this condition works out to be

$$\Delta x = \sqrt{\frac{x(1-x)}{N(x)}} d_{min}, \tag{A3}$$

where Δx is the separation between successive signals. The density $\rho(x)$ of signals is therefore

$$\rho(x) = \frac{1}{\Delta x} = \sqrt{\frac{N(x)}{x(1-x)}} \frac{1}{d_{min}}. \tag{A4}$$

Alice wants to maximize the number of distinct signals. So, in choosing the function $N(x)$, she needs to solve the following optimization problem: maximize the quantity (from Equation (A4))

$$\int_0^1 \sqrt{\frac{N(x)}{x(1-x)}} dx, \tag{A5}$$

while satisfying the following two constraints (which come from Equations (A1) and (A2), combined with Equation (A4))

$$\int_0^1 \frac{N(x)^{3/2} - \mathcal{N} N(x)^{1/2}}{\sqrt{x(1-x)}} dx = 0. \tag{A6}$$

$$\int_0^1 \frac{x N(x)^{3/2} - \mathcal{N}_H N(x)^{1/2}}{\sqrt{x(1-x)}} dx = 0. \tag{A7}$$

This problem can be solved by the calculus of variations, and it can be found that Alice should choose $N(x)$ to be of the form

$$N(x) \propto \frac{1}{\frac{x}{\lambda} + \frac{1-x}{1-\lambda}}. \tag{A8}$$

Here, λ is a real number between zero and one, fixed by the requirement that the overall probability of heads must equal $\mathcal{N}_H/\mathcal{N}$ (we could have written the result in other ways; we use λ only to facilitate our later comparison with the Scrooge distribution). Once the value of λ is set, the constant factor multiplying the right-hand side is fixed by Equation (A6).

We now use this result to generate the probability distribution $\hat{\sigma}(x)$ over the probability simplex. We define it as follows: in many rounds of communication, we want $\hat{\sigma}(x)dx$ to approximate the fraction of the total number of tosses that come from a coin whose probability of heads is between x and $x + dx$. More precisely, we define $\hat{\sigma}(x)$ to be proportional to $N(x)\rho(x)$, with the proportionality constant set by the normalization condition $\int_0^1 \hat{\sigma}(x)dx = 1$ (we have multiplied $\rho(x)$ by $N(x)$ to turn the density of signals into the density of tosses). By substituting for $N(x)$ and $\rho(x)$ in accordance with Equations (A4) and (A8), we arrive at

$$\hat{\sigma}(x) = \frac{A}{\sqrt{x(1-x)}} \cdot \frac{1}{\left(\frac{x}{\lambda} + \frac{1-x}{1-\lambda}\right)^{3/2}}, \tag{A9}$$

where A is the normalization constant. Comparing this form with that of Equation (25), we see that this alternative problem does not lead us to the real-amplitude Scrooge distribution—the exponent appearing in the denominator is 3/2 instead of 2. Moreover, λ and $1 - \lambda$ have no obvious meaning in this problem, whereas in the problem considered in Sections 4 and 5, the λ_js can be interpreted directly in terms of the imposed bounds $\mathcal{M}_j$ on the expectation values of the number of times that the various outcomes occur.

References

1. Holevo, A.S. The Capacity of Quantum Channel with General Signal States. *IEEE Trans. Inf. Theory* **1998**, *44*, 269–273. [CrossRef]
2. Schumacher, B.; Westmoreland, M. Sending Classical Information via Noisy Quantum Channels. *Phys. Rev. A* **1997**, *56*, 131–138. [CrossRef]
3. Holevo, A.S. Bounds for the quantity of information transmitted by a quantum communication channel. *Prob. Inf. Transm.* **1973**, *9*, 177–183.
4. Jozsa, R.; Robb, D.; Wootters, W.K. Lower bound for accessible information in quantum mechanics. *Phys. Rev. A* **1994**, *49*, 668–677. [CrossRef] [PubMed]
5. Goldstein, S.; Lebowitz, J.L.; Tumulka, R.; Zanghì, N. On the distribution of the wave function for systems in thermal equilibrium. *J. Stat. Phys.* **2006**, *125*, 1193–1221. [CrossRef]
6. Tumulka, R.; Zanghì, N. Smoothness of wave functions in thermal equilibrium. *J. Math. Phys.* **2005**, *46*, 112104. [CrossRef]
7. Reimann, P. Typicality of pure states randomly sampled according to the Gaussian adjusted projected measure. *J. Stat. Phys.* **2008**, *132*, 921–935. [CrossRef]
8. Goldstein, S.; Lebowitz, J.L.; Mastrodonato, C.; Tumulka, R.; Zanghì, N. Universal probability distribution for the wave function of a quantum system entangled with its environment. *Commun. Math. Phys.* **2016**, *342*, 965–988. [CrossRef]

9. Sen(De), A.; Sen, U.; Lewenstein, M. Distillation Protocols that Involve Local Distinguishing: Composing Upper and Lower Bounds on Locally Accessible Information. *Phys. Rev. A* **2006**, *74*, 052332. [CrossRef]
10. Wootters, W.K. Random Quantum States. *Found. Phys.* **1990** *20*, 1365–1378. [CrossRef]
11. Jones, K.R.W. Riemann-Liouville fractional integration and reduced distributions on hyperspheres. *J. Phys. A Math. Gen.* **1991**, *24*, 1237–1244. [CrossRef]
12. Jacobs, K. On the properties of information gathering in quantum and classical measurements. *arXiv* **2003**, arxiv:quant-ph/0304200.
13. Jacobs, K. Efficient measurements, purification, and bounds on the mutual information. *Phys. Rev. A* **2003**, *68*, 054302. [CrossRef]
14. Dall'Arno, M. Hierarchy of Bounds on Accessible Information and Informational Power. *Phys. Rev. A* **2015**, *92*, 012328. [CrossRef]
15. Mintert, F.; Zyczkowski, K. Wehrl entropy, Lieb conjecture and entanglement monotones. *Phys. Rev. A* **2004**, *69*, 022317. [CrossRef]
16. Cheng, S.; Hall, M.J.W. Complementarily relations for quantum coherence. *Phys. Rev. A* **2015**, *92*, 042101. [CrossRef]
17. Zhang, L.; Singh, U.; Pati, A.K. Average subentropy, coherence and entanglement of random mixed quantum states. *Ann. Phys.* **2017**, *377*, 125–146. [CrossRef]
18. Zhang, L. Average coherence and its typicality for random mixed quantum states. *J. Phys. A Math. Theor.* **2017**, *50*, 155303. [CrossRef]
19. Zhang, L.; Ma, Z.; Chen, Z.; Fei, S.-M. Coherence generating power of unitary transformations via probabilistic average. *Quantum Inf. Process.* **2018**, *17*, 186. [CrossRef]
20. Nichols, S.R.; Wootters, W.K. Between entropy and subentropy. *Quantum Inf. Comput.* **2003**, *3*, 1–14.
21. Mitchison, G.; Jozsa, R. Towards a geometric interpretation of quantum information compression. *Phys. Rev. A* **2004**, *69*, 032304. [CrossRef]
22. Datta, N.; Dorlas, T.; Jozsa, R.; Benatti, F. Properties of subentropy. *J. Math. Phys.* **2014**, *55*, 062203. [CrossRef]
23. Jozsa, R.; Mitchison, G. Symmetric polynomials in information theory: Entropy and subentropy. *J. Math. Phys.* **2015**, *56*, 062201. [CrossRef]
24. Audenaert, K.; Datta, N.; Ozols, M. Entropy power inequalities for qubits. *J. Math. Phys.* **2016**, *57*, 052202. [CrossRef]
25. Sýkora, S. Quantum theory and the bayesian inference problems. *J. Stat. Phys.* **1974**, *11*, 17–27. [CrossRef]
26. Wootters, W.K. Communicating through Probabilities: Does Quantum Theory Optimize the Transfer of Information? *Entropy* **2013**, *15*, 3130–3147. [CrossRef]
27. Wootters, W.K. Statistical distance and Hilbert space. *Phys. Rev. D* **1981**, *23*, 357. [CrossRef]
28. Wootters, W.K. Optimal information transfer and real-vector-space quantum theory. In *Quantum Theory: Informational Foundations and Foils*; Chiribella, G., Spekkens, R.W., Eds.; Springer: Dordrecht, The Netherlands, 2016; pp. 21–43.

© 2018 by the author. Licensee MDPI, Basel, Switzerland. This article is an open access article distributed under the terms and conditions of the Creative Commons Attribution (CC BY) license (http://creativecommons.org/licenses/by/4.0/).

Article

Attacks against a Simplified Experimentally Feasible Semiquantum Key Distribution Protocol

Michel Boyer [1], Rotem Liss [2,*] and Tal Mor [2]

[1] Département d'Informatique et de Recherche Opérationnelle (DIRO), Université de Montréal, Montréal, QC H3C 3J7, Canada; boyer@iro.umontreal.ca
[2] Computer Science Department, Technion, Haifa 3200003, Israel; talmo@cs.technion.ac.il
* Correspondence: rotemliss@cs.technion.ac.il; Tel.: +972-4-829-3826

Received: 16 June 2018; Accepted: 16 July 2018; Published: 18 July 2018

Abstract: A semiquantum key distribution (SQKD) protocol makes it possible for a quantum party and a classical party to generate a secret shared key. However, many existing SQKD protocols are not experimentally feasible in a secure way using current technology. An experimentally feasible SQKD protocol, "classical Alice with a controllable mirror" (the "Mirror protocol"), has recently been presented and proved completely robust, but it is more complicated than other SQKD protocols. Here we prove a simpler variant of the Mirror protocol (the "simplified Mirror protocol") to be completely non-robust by presenting two possible attacks against it. Our results show that the complexity of the Mirror protocol is at least partly necessary for achieving robustness.

Keywords: quantum key distribution; semiquantum key distribution; security; attack

1. Introduction

Quantum key distribution (QKD) protocols allow two parties, Alice and Bob, to share a secret random key that is secure even against the most powerful adversaries. Semiquantum key distribution (SQKD) protocols achieve the same goal even if one of the two parties (Alice or Bob) is limited to use only classical operations: the classical party can use only the computational basis $\{|0\rangle, |1\rangle\}$, while the quantum party can use any basis—for example, both the computational basis and the Hadamard basis $\{|+\rangle \triangleq \frac{|0\rangle+|1\rangle}{\sqrt{2}}, |-\rangle \triangleq \frac{|0\rangle-|1\rangle}{\sqrt{2}}\}$. As explained in [1,2], the importance of SQKD protocols is both conceptual and practical: they make it possible to investigate the amount of "quantumness" needed for QKD, and they may, in some cases, be easier to implement than standard QKD protocols.

The first SQKD protocol was "QKD with classical Bob" [1]. Later, other SQKD protocols have been suggested, including "QKD with classical Alice" [3,4] and many others (e.g., [2,5–9]). Most SQKD protocols have been proven "robust": namely [1], if the adversary Eve succeeds in getting some secret information, she must cause some errors that may be noticed by Alice and Bob. A few SQKD protocols also have a security analysis [10–13]. Proving robustness is a step towards proving security; proving the security of SQKD protocols is difficult because those protocols are usually two-way: for example, Alice sends a quantum state to Bob, and Bob performs a specific classical operation and sends the resulting quantum state back to Alice.

However, many SQKD protocols, including [1,3], are vulnerable to practical attacks and cannot be experimentally constructed in a secure way using current technology. An important classical operation of those protocols is named SIFT. The definition of a SIFT operation performed by Alice (assuming that Alice is the classical party) is as follows: Alice measures the incoming quantum state in the computational basis $\{|0\rangle, |1\rangle\}$ and then generates the state she measured and resends it towards Bob. Security of those SQKD protocols relies on the assumption that during the SIFT operation, Alice's measurement devices can measure the *precise* states $\{|0\rangle, |1\rangle\}$ and distinguish those precise states from any imperfect similar state, and Alice's photon generation devices can generate the *precise* states $\{|0\rangle, |1\rangle\}$ and not any other (imperfect) state. In particular, the generated states $\{|0\rangle, |1\rangle\}$ must be indistinguishable from states that Alice *reflects* towards Bob. Using current photonic technology, Alice's devices are imperfect, which makes this assumption incorrect and makes possible attacks by the eavesdropper Eve: for example, Eve may send a slightly modified state towards Alice (a "tagging" attack") or may distinguish between the states sent by Alice. Full details about those practical attacks are available in [14–16].

An experimentally feasible SQKD protocol named "classical Alice with a controllable mirror" (the "Mirror protocol") has recently been presented [16]. This protocol is safe against the "tagging" attack presented by [14]. Moreover, the protocol was proved by [16] to be completely robust against any attacker Eve, even if Eve is all-powerful and limited only by the laws of physics, and even if Eve can send multi-photon pulses. The robustness proof is still correct even if the detectors of Alice and Bob cannot *count* how many photons arrive in each mode: namely, when either Alice or Bob looks at a detector, which detects a specific mode, they can only notice whether it "clicks" (detects one photon or more in that mode) or not (finds the mode to be empty). This is the standard situation when using current technology.

In this paper, we present a simpler variant of the Mirror protocol (the "simplified Mirror protocol"), which is easier to implement. Our variant allows the classical party, Alice, to choose one of three operations, while the Mirror protocol allows Alice to choose one of four operations. We present two attacks against this variant, proving it to be non-robust. Our results show that the four classical operations allowed by the Mirror protocol are probably necessary for robustness.

In Section 2 we present the Mirror protocol described by [16]. In Section 3 we present the simplified Mirror protocol and its motivation. In Section 4 we prove the simplified Mirror protocol to be non-robust by presenting two attacks against it: a full attack and a weaker attack. In Section 5 we discuss potential implications of our results.

2. The Mirror Protocol

For describing the Mirror protocol (presented by [16]), we assume a photonic implementation consisting of two modes: the mode of the qubit state $|0\rangle$ and the mode of the qubit state $|1\rangle$ (below we call them "the $|0\rangle$ mode" and "the $|1\rangle$ mode", respectively). For example, the $|0\rangle$ mode and the $|1\rangle$ mode can represent two different polarizations or two different time bins. We use the Fock space notations: if there is exactly one photon (and, thus, our Hilbert space is the qubit space), the Fock state $|0,1\rangle$ (equivalent to $|0\rangle$) represents one photon in the $|0\rangle$ mode, and the Fock state $|1,0\rangle$ (equivalent to $|1\rangle$) represents one photon in the $|1\rangle$ mode. We can extend the qubit space to a 3-dimensional Hilbert space by adding the Fock "vacuum state" $|0,0\rangle$, which represents an absence of photons. Most generally, the Fock state $|m_1, m_0\rangle$ represents m_1 indistinguishable photons in the $|1\rangle$ mode and m_0 indistinguishable photons in the $|0\rangle$ mode. Similarly (in the Hadamard basis), the Fock state $|m_-, m_+\rangle_x$ represents m_- indistinguishable photons in the $|-\rangle$ mode and m_+ indistinguishable photons in the $|+\rangle$ mode. More details about the Fock space notations are given in [16]; it is vital to use those mathematical notations for describing and analyzing all practical attacks on a QKD protocol (see [17] for details).

In the Mirror protocol, in each round, Bob sends to Alice the $|+\rangle_B$ state—namely, the $|0,1\rangle_{x,B} \triangleq \frac{|0,1\rangle_B + |1,0\rangle_B}{\sqrt{2}}$ state. Then, Alice prepares an ancillary state in the initial vacuum state $|0,0\rangle_A$ and chooses at random one of the following four classical operations:

- **I (CTRL)** Reflect all the photons towards Bob, without measuring any photon. The mathematical description is:

$$I |0,0\rangle_A |m_1, m_0\rangle_B = |0,0\rangle_A |m_1, m_0\rangle_B. \qquad (1)$$

- S_1 (**SWAP-10**) Reflect all photons in the $|o\rangle$ mode towards Bob, and measure all photons in the $|1\rangle$ mode. The mathematical description is:

$$S_1 |0,0\rangle_A |m_1, m_o\rangle_B = |m_1, 0\rangle_A |0, m_o\rangle_B. \qquad (2)$$

- S_0 (**SWAP-01**) Reflect all photons in the $|1\rangle$ mode towards Bob, and measure all photons in the $|o\rangle$ mode. The mathematical description is:

$$S_0 |0,0\rangle_A |m_1, m_o\rangle_B = |0, m_o\rangle_A |m_1, 0\rangle_B. \qquad (3)$$

- **S** (**SWAP-ALL**) Measure all the photons, without reflecting any photon towards Bob. The mathematical description is:

$$S |0,0\rangle_A |m_1, m_o\rangle_B = |m_1, m_o\rangle_A |0, 0\rangle_B. \qquad (4)$$

(We note that in the above mathematical description, Alice measures her ancillary state $|\cdot\rangle_A$ in the computational basis and sends back to Bob the $|\cdot\rangle_B$ state.)

The states sent from Alice to Bob (without any error, loss, or eavesdropping) are detailed in Table 1.

Table 1. The state sent from Alice to Bob in the Mirror protocol without errors or losses, depending on Alice's classical operation and on whether Alice detected a photon or not.

Alice's Classical Operation	Did Alice Detect a Photon?	State Sent from Alice to Bob
CTRL	no (happens with certainty)	$|0,1\rangle_{x,B} = \frac{1}{\sqrt{2}}[|0,1\rangle_B + |1,0\rangle_B]$
SWAP-10	no (happens with probability $\frac{1}{2}$)	$|0,1\rangle_B$
SWAP-10	yes (happens with probability $\frac{1}{2}$)	$|0,0\rangle_B$
SWAP-01	no (happens with probability $\frac{1}{2}$)	$|1,0\rangle_B$
SWAP-01	yes (happens with probability $\frac{1}{2}$)	$|0,0\rangle_B$
SWAP-ALL	yes (happens with certainty)	$|0,0\rangle_B$

Then, Bob measures the incoming state in a random basis (either the computational basis $\{|0\rangle, |1\rangle\}$ or the Hadamard basis $\{|+\rangle, |-\rangle\}$). After completing all rounds, Alice sends over the classical channel her operation choices (CTRL, SWAP-x, or SWAP-ALL; she keeps $x \in \{01, 10\}$ in secret), Bob sends over the classical channel his basis choices, and both of them reveal some non-secret information on their measurement results (as elaborated in [16]). Then, Alice and Bob reveal and compute the error rate on test bits for which Alice used SWAP-10 or SWAP-01 and Bob measured in the computational basis, and the error rate on test bits for which Alice used CTRL and Bob measured in the Hadamard basis. They also check whether other errors exist (for example, they verify Bob detects no photons in case Alice uses SWAP-ALL). Alice and Bob also discard mismatched rounds, such as rounds in which Alice used SWAP-10 and Bob used the Hadamard basis. Alice and Bob share the secret bit 0 if Alice uses SWAP-10 and detects no photon while Bob measures in the computational basis and detects a photon in the $|o\rangle$ mode; similarly, they share the secret bit 1 if Alice uses SWAP-01 and detects no photon while Bob measures in the computational basis and detects a photon in the $|1\rangle$ mode.

Finally, Alice and Bob verify that the error rates are below some thresholds, and they perform error correction and privacy amplification in the standard way for QKD protocols. At the end of the protocol, Alice and Bob hold an identical final key that is completely secure against any eavesdropper.

A full description of the protocol and a proof of its complete robustness are both available in [16].

The experimental implementation of the protocol can use two time bins (namely, two pulses), one for the $|o\rangle$ mode and one for the $|1\rangle$ mode. In this case, Alice's possible operations can be described as possible ways for operating a controllable mirror, so that Alice can choose whether to reflect or measure the photon(s) in each time bin. The mirror can be experimentally implemented in various ways; for example:

- It can be implemented as a mechanically moved mirror. Such mirror is trivial to implement, but it is very slow.
- It can be implemented by using optical elements: an electronically-triggered Pockels cell, which changes the polarization of the photon(s) in one of the pulses, and a polarizing beam splitter, which can split the two different pulses (that now have different polarizations) into two paths. This implementation is feasible and gives much higher bit rates than the mechanical implementation.

More details about the experimental implementations are available in [16].

3. The "Simplified Mirror Protocol": A Simpler and Non-Robust Variant of the Mirror Protocol

In this paper, we discuss a simpler variant of the Mirror protocol, which we name the "simplified Mirror protocol". The simplified Mirror protocol is identical to the Mirror protocol described in Section 2, except that it does not include the SWAP-ALL operation. In other words, in the simplified protocol, Alice chooses at random one of the three classical operations CTRL, SWAP-10, and SWAP-01.

The simplified protocol is easier to implement, because the SWAP-ALL operation poses some experimental challenges to the electronic implementation discussed in Section 2: for implementing SWAP-ALL, the Pockels cell should either remain working for a long time (changing polarization for both time bins) or be operated twice (changing polarization for each time bin separately). In more details, for the two pulses representing the $|o\rangle$ mode and the $|1\rangle$ mode: if we assume the duration of each pulse is t and the time difference between the two pulses is T (where $t \ll T$), the first solution means keeping the Pockels cell operating during the time period $[0, T + 2t]$, and the second solution means operating the Pockels cell during the two time periods $[0, t]$ and $[T + t, T + 2t]$. The first solution may be problematic for some models of the Pockels cell, and the second solution may be problematic because of the recovery time needed for the Pockels cell. Therefore, at least in some implementations, the simplified Mirror protocol is much easier to implement than the standard Mirror protocol.

Moreover, analyzing the simplified protocol gives a better understanding of the properties required for an SQKD protocol to be robust. In particular, this analysis explains why the structure and complexity of the Mirror protocol are necessary for robustness.

For completeness, we provide below the full description of the simplified Mirror protocol. We note that this description is almost the same as the description of the Mirror protocol in Section 2.

In the simplified Mirror protocol, in each round, Bob sends to Alice the $|+\rangle_B$ state—namely, the $|0,1\rangle_{x,B} \triangleq \frac{|0,1\rangle_B + |1,0\rangle_B}{\sqrt{2}}$ state. Then, Alice prepares an ancillary state in the initial vacuum state $|0,0\rangle_A$ and chooses at random one of the following three classical operations:

- **I (CTRL)** Reflect all the photons towards Bob, without measuring any photon. The mathematical description is:

$$I |0,0\rangle_A |m_1, m_o\rangle_B = |0,0\rangle_A |m_1, m_o\rangle_B. \tag{5}$$

- **S_1 (SWAP-10)** Reflect all photons in the $|o\rangle$ mode towards Bob, and measure all photons in the $|1\rangle$ mode. The mathematical description is:

$$S_1 |0,0\rangle_A |m_1, m_o\rangle_B = |m_1, 0\rangle_A |0, m_o\rangle_B. \tag{6}$$

- **S_0 (SWAP-01)** Reflect all photons in the $|1\rangle$ mode towards Bob, and measure all photons in the $|o\rangle$ mode. The mathematical description is:

$$S_0 |0,0\rangle_A |m_1, m_o\rangle_B = |0, m_o\rangle_A |m_1, 0\rangle_B. \tag{7}$$

(We note that in the above mathematical description, Alice measures her ancillary state $|\cdot\rangle_A$ in the computational basis and sends back to Bob the $|\cdot\rangle_B$ state.)

The states sent from Alice to Bob (without any error, loss, or eavesdropping) are detailed in Table 2.

Table 2. The state sent from Alice to Bob in the simplified Mirror protocol without errors or losses, depending on Alice's classical operation and on whether Alice detected a photon or not.

Alice's Classical Operation	Did Alice Detect a Photon?	State Sent from Alice to Bob
CTRL	no (happens with certainty)	$\|0,1\rangle_{x,B} = \frac{1}{\sqrt{2}}[\,\|0,1\rangle_B + \|1,0\rangle_B\,]$
SWAP-10	no (happens with probability $\frac{1}{2}$)	$\|0,1\rangle_B$
SWAP-10	yes (happens with probability $\frac{1}{2}$)	$\|0,0\rangle_B$
SWAP-01	no (happens with probability $\frac{1}{2}$)	$\|1,0\rangle_B$
SWAP-01	yes (happens with probability $\frac{1}{2}$)	$\|0,0\rangle_B$

Then, Bob measures the incoming state in a random basis (either the computational basis $\{\,|0\rangle, |1\rangle\,\}$ or the Hadamard basis $\{\,|+\rangle, |-\rangle\,\}$). After completing all rounds, Alice sends over the classical channel her operation choices (CTRL or SWAP-x; she keeps $x \in \{01, 10\}$ in secret), Bob sends over the classical channel his basis choices, and both of them reveal some non-secret information on their measurement results (as elaborated in [16]). Then, Alice and Bob reveal and compute the error rate on test bits for which Alice used SWAP-10 or SWAP-01 and Bob measured in the computational basis, and the error rate on test bits for which Alice used CTRL and Bob measured in the Hadamard basis. They also check whether other errors exist (for example, it must never happen that *both* Alice and Bob detect a photon). Alice and Bob also discard mismatched rounds, such as rounds in which Alice used SWAP-10 and Bob used the Hadamard basis. Alice and Bob share the secret bit 0 if Alice uses SWAP-10 and detects no photon while Bob measures in the computational basis and detects a photon in the $|0\rangle$ mode; similarly, they share the secret bit 1 if Alice uses SWAP-01 and detects no photon while Bob measures in the computational basis and detects a photon in the $|1\rangle$ mode.

Finally, Alice and Bob verify that the error rates are below some thresholds, and they perform error correction and privacy amplification in the standard way for QKD protocols. At the end of the protocol, Alice and Bob hold an identical final key that is completely secure against any eavesdropper.

4. Attacks against the Simplified Mirror Protocol

We prove the simplified protocol to be non-robust by presenting two attacks: a "full attack" described in Section 4.1, which gives Eve full information but causes full loss of the CTRL bits, and a "weaker attack" described in Section 4.2, which gives Eve less information but causes fewer losses of CTRL bits.

4.1. A Full Attack on the Simplified Protocol that Gives Eve Full Information

In this attack, Eve gets full information of all the information bits. Namely, she gets full information on the SWAP-10 and SWAP-01 bits that were measured by Bob in the computational basis.

Eve applies her attack in two stages: the first stage is on the way from Bob to Alice, and the second stage is on the way from Alice to Bob. In both stages she uses her own probe space (namely, ancillary space) $\mathcal{H}_E = \mathcal{H}_3$ spanned by the orthonormal basis $\{\,|0\rangle_E, |1\rangle_E, |2\rangle_E\,\}$. We assume that Eve fully controls the environment, the errors, and the losses (this is a standard assumption when analyzing the security of QKD): namely, no losses and no errors exist between Bob and Eve or between Alice and Eve.

In the first stage of the attack (on the way from Bob to Alice), Eve intercepts the state $|+\rangle_B$ (namely, $|0,1\rangle_{x,B}$) sent by Bob, generates instead the state

$$\frac{1}{\sqrt{3}}[|0,1\rangle_B |1\rangle_E + |1,0\rangle_B |1\rangle_E + |0,0\rangle_B |0\rangle_E] = \sqrt{\frac{2}{3}} |0,1\rangle_{x,B} |1\rangle_E + \sqrt{\frac{1}{3}} |0,0\rangle_B |0\rangle_E, \quad (8)$$

and sends to Alice the B part of the state. This state causes Alice to get no photons with probability $\frac{1}{3}$ and get the expected $|+\rangle_B$ state with probability $\frac{2}{3}$. Alice then performs at random one of the three classical operations CTRL, SWAP-10, or SWAP-01. The resulting possible states of Bob+Eve are described in Table 3.

Table 3. The state of Bob+Eve after Alice's classical operation for the attacks described in Sections 4.1 and 4.2, depending on Alice's classical operation and on whether Alice detected a photon or not.

Alice's Classical Operation	Did Alice Detect a Photon?	Bob+Eve State
CTRL	no (happens with certainty)	$\frac{1}{\sqrt{3}}[|0,1\rangle_B |1\rangle_E + |1,0\rangle_B |1\rangle_E + |0,0\rangle_B |0\rangle_E]$
SWAP-10	no (happens with probability $\frac{2}{3}$)	$\frac{1}{\sqrt{2}}[|0,1\rangle_B |1\rangle_E + |0,0\rangle_B |0\rangle_E]$
SWAP-10	yes (happens with probability $\frac{1}{3}$)	$|0,0\rangle_B |1\rangle_E$
SWAP-01	no (happens with probability $\frac{2}{3}$)	$\frac{1}{\sqrt{2}}[|1,0\rangle_B |1\rangle_E + |0,0\rangle_B |0\rangle_E]$
SWAP-01	yes (happens with probability $\frac{1}{3}$)	$|0,0\rangle_B |1\rangle_E$

In the second stage of the attack (on the way from Alice to Bob), Eve applies the unitary operator V on the joint Bob+Eve state, where V is defined as follows:

$$V|0,1\rangle_B |1\rangle_E = -\sqrt{\frac{1}{3}} |1,0\rangle_B |1\rangle_E + \sqrt{\frac{2}{3}} |0,0\rangle_B |0\rangle_E, \quad (9)$$

$$V|1,0\rangle_B |1\rangle_E = -\sqrt{\frac{1}{3}} |0,1\rangle_B |0\rangle_E + \sqrt{\frac{2}{3}} |0,0\rangle_B |1\rangle_E, \quad (10)$$

$$V|0,0\rangle_B |0\rangle_E = \sqrt{\frac{1}{3}} |0,1\rangle_B |0\rangle_E + \sqrt{\frac{1}{3}} |1,0\rangle_B |1\rangle_E + \sqrt{\frac{1}{3}} |0,0\rangle_B |+\rangle_E, \quad (11)$$

$$V|0,0\rangle_B |1\rangle_E = |0,0\rangle_B |2\rangle_E. \quad (12)$$

V is indeed a unitary operator, because we can prove the right-hand sides to be orthonormal: all the right-hand sides are normalized vectors; the first two vectors are clearly orthogonal; the third vector is orthogonal to the first two, because $\langle 0|+\rangle_E = \langle 1|+\rangle_E = \frac{1}{\sqrt{2}}$; and the fourth vector is orthogonal to the three others. Thus, V defines (or, more precisely, can be extended to) a unitary operator on $\mathcal{H}_B \otimes \mathcal{H}_E$.

Applying the unitary operator V on Table 3 gives the states listed in Table 4. Comparing it with Table 2, we conclude that this attack never causes Alice and Bob to detect an error. Moreover, Eve detects the whole secret key: Eve measures "0" in her probe if Alice and Bob agree on the bit 0, and she measures "1" in her probe if Alice and Bob agree on the bit 1. However, Eve causes several kinds of losses; in particular, all the CTRL bits are lost.

Therefore, this attack makes it possible for Eve to get full information without inducing any error. However, Eve causes many losses, including full loss of the CTRL bits.

Table 4. The state of Bob+Eve after completing Eve's attack described in Section 4.1, depending on Alice's classical operation and on whether Alice detected a photon or not.

Alice's Classical Operation	Did Alice Detect a Photon?	Bob+Eve State
CTRL	no (happens with certainty)	$\lvert 0,0\rangle_B \lvert +\rangle_E$
SWAP-10	no (happens with probability $\frac{2}{3}$)	$\frac{1}{\sqrt{6}}\lvert 0,1\rangle_B \lvert 0\rangle_E + \lvert 0,0\rangle_B \frac{3\lvert 0\rangle_E + \lvert 1\rangle_E}{\sqrt{12}}$
SWAP-10	yes (happens with probability $\frac{1}{3}$)	$\lvert 0,0\rangle_B \lvert 2\rangle_E$
SWAP-01	no (happens with probability $\frac{2}{3}$)	$\frac{1}{\sqrt{6}}\lvert 1,0\rangle_B \lvert 1\rangle_E + \lvert 0,0\rangle_B \frac{\lvert 0\rangle_E + 3\lvert 1\rangle_E}{\sqrt{12}}$
SWAP-01	yes (happens with probability $\frac{1}{3}$)	$\lvert 0,0\rangle_B \lvert 2\rangle_E$

4.2. A Weaker Attack on the Simplified Protocol Causing Fewer Losses of the CTRL Bits

The full attack described in Section 4.1 makes it impossible for Bob to ever detect a CTRL bit, which may look suspicious. We now present a weaker attack that lets Bob detect some CTRL bits but gives Eve less information.

The first stage of the attack (on the way from Bob to Alice) remains the same: that is, the state Eve sends to Alice is still given by Equation (8), and the resulting Bob+Eve state after Alice's classical operation is still shown in Table 3. Eve's probe space is, too, the same as before: $\mathcal{H}_E = \mathcal{H}_3 \triangleq \text{Span}\{\lvert 0\rangle_E, \lvert 1\rangle_E, \lvert 2\rangle_E\}$.

This attack is characterized by the parameter $0 \leq \epsilon \leq 1$. We will see that $\epsilon = 0$ gives the full attack described in Section 4.1, while $\epsilon = 1$ gives Eve no information at all.

Another important parameter used by the attack is

$$\kappa \triangleq \sqrt{\frac{1-\epsilon^2}{3-2\epsilon^2}}. \tag{13}$$

We notice that for small values of ϵ, the value of κ is close to $\sqrt{\frac{1}{3}}$. Moreover, for all $0 \leq \epsilon \leq 1$, it holds that $0 < \epsilon^2 + \kappa^2 \leq 1$ and $2\kappa^2 < 1$.

In the second stage of the attack (on the way from Alice to Bob), Eve applies the unitary operator V on the joint Bob+Eve state, where V is defined as follows:

$$V\lvert 0,1\rangle_B \lvert 1\rangle_E = \epsilon\lvert 0,1\rangle_B\lvert 2\rangle_E - \kappa\lvert 1,0\rangle_B\lvert 1\rangle_E + \sqrt{1-\kappa^2-\epsilon^2}\lvert 0,0\rangle_B\lvert 0\rangle_E, \tag{14}$$
$$V\lvert 1,0\rangle_B \lvert 1\rangle_E = -\kappa\lvert 0,1\rangle_B\lvert 0\rangle_E + \epsilon\lvert 1,0\rangle_B\lvert 2\rangle_E + \sqrt{1-\kappa^2-\epsilon^2}\lvert 0,0\rangle_B\lvert 1\rangle_E, \tag{15}$$
$$V\lvert 0,0\rangle_B \lvert 0\rangle_E = \kappa\lvert 0,1\rangle_B\lvert 0\rangle_E + \kappa\lvert 1,0\rangle_B\lvert 1\rangle_E + \sqrt{1-2\kappa^2}\lvert 0,0\rangle_B\lvert +\rangle_E, \tag{16}$$
$$V\lvert 0,0\rangle_B \lvert 1\rangle_E = \lvert 0,0\rangle_B\lvert 2\rangle_E. \tag{17}$$

V is indeed a unitary operator, because we can prove the right-hand sides to be orthonormal: all the right-hand sides are clearly normalized; the first two vectors are orthogonal; the fourth vector is orthogonal to the three others; and the third vector is orthogonal to the first and to the second, because

$$1 - 2\kappa^2 = \frac{3 - 2\epsilon^2 - 2(1-\epsilon^2)}{3 - 2\epsilon^2} = \frac{1}{3 - 2\epsilon^2}, \tag{18}$$

$$1 - \kappa^2 - \epsilon^2 = \frac{(3-2\epsilon^2) - (1-\epsilon^2) - (3\epsilon^2 - 2\epsilon^4)}{3 - 2\epsilon^2} = \frac{2(1-\epsilon^2)^2}{3 - 2\epsilon^2}, \tag{19}$$

and thus $\frac{\sqrt{1-\kappa^2-\epsilon^2}\sqrt{1-2\kappa^2}}{\sqrt{2}} = \kappa^2$. Therefore, V extends to a unitary operator on $\mathcal{H}_B \otimes \mathcal{H}_E$.

The final global state after Eve's attack is described in Table 5 (calculated by applying the operator V on Table 3), given the following definitions:

$$a \triangleq \sqrt{1 - \kappa^2 - \epsilon^2} + \frac{\sqrt{1 - 2\kappa^2}}{\sqrt{2}}, \tag{20}$$

$$b \triangleq \frac{\sqrt{1 - 2\kappa^2}}{\sqrt{2}}. \tag{21}$$

Table 5. The state of Bob+Eve after completing Eve's attack described in Section 4.2, depending on Alice's classical operation and on whether Alice detected a photon or not. The parameters a and b are defined in Equations (20) and (21).

Alice's Classical Operation	Did Alice Detect a Photon?	Bob+Eve State
CTRL	no (happens with certainty)	$\sqrt{\frac{2\epsilon^2}{3}} \lvert 0,1 \rangle_{x,B} \lvert 2 \rangle_E + \sqrt{1 - \frac{2\epsilon^2}{3}} \lvert 0,0 \rangle_B \lvert + \rangle_E$
SWAP-10	no (happens with probability $\frac{2}{3}$)	$\frac{1}{\sqrt{2}} [\lvert 0,1 \rangle_B (\epsilon \lvert 2 \rangle_E + \kappa \lvert 0 \rangle_E) + \lvert 0,0 \rangle_B (a \lvert 0 \rangle_E + b \lvert 1 \rangle_E)]$
SWAP-10	yes (happens with probability $\frac{1}{3}$)	$\lvert 0,0 \rangle_B \lvert 2 \rangle_E$
SWAP-01	no (happens with probability $\frac{2}{3}$)	$\frac{1}{\sqrt{2}} [\lvert 1,0 \rangle_B (\epsilon \lvert 2 \rangle_E + \kappa \lvert 1 \rangle_E) + \lvert 0,0 \rangle_B (b \lvert 0 \rangle_E + a \lvert 1 \rangle_E)]$
SWAP-01	yes (happens with probability $\frac{1}{3}$)	$\lvert 0,0 \rangle_B \lvert 2 \rangle_E$

We notice that for $\epsilon = 0$, the attack is the same as in Section 4.1. If $\epsilon = 1$, the loss rate of CTRL bits is $\frac{1}{3}$, and Eve gets no information at all on the information bits (because $\kappa = 0$).

In general, if Alice and Bob share a "secret" bit $b \in \{0,1\}$, Eve's probe state is in the (normalized) state

$$\frac{\epsilon \lvert 2 \rangle_E + \kappa \lvert b \rangle_E}{\sqrt{\epsilon^2 + \kappa^2}}. \tag{22}$$

When Eve measures her probe state in the computational basis $\{\lvert 0 \rangle_E, \lvert 1 \rangle_E, \lvert 2 \rangle_E\}$, she gets the information bit b with probability

$$p = \frac{\kappa^2}{\epsilon^2 + \kappa^2} = \frac{1 - \epsilon^2}{1 + 2\epsilon^2 - 2\epsilon^4}, \tag{23}$$

and the loss rates of CTRL and SWAP-x bits (where $x \in \{01, 10\}$) are

$$R_{\text{CTRL}} = 1 - \frac{2\epsilon^2}{3}, \tag{24}$$

$$R_{\text{SWAP-}x} = 1 - \frac{\epsilon^2 + \kappa^2}{2}, \tag{25}$$

respectively.

Table 6 shows the probabilities p and the loss rates $R_{\text{CTRL}}, R_{\text{SWAP-}x}$ for various values of ϵ. For example, for $\epsilon = 0.5$, Eve still gets the information bit with probability $p \approx 0.55$, Bob's loss rate for the CTRL bits is $R_{\text{CTRL}} \approx 0.83$, and his loss rate for the SWAP-x bits is $R_{\text{SWAP-}x} \approx 0.73$.

Table 6. The probability p of Eve obtaining an information bit, and the loss rates R_{CTRL} and $R_{\text{SWAP-}x}$ of CTRL and SWAP-x bits (where $x \in \{01, 10\}$), respectively, for several values of the attack's parameter ϵ.

ϵ	0	0.1	0.2	0.3	0.4	0.5	0.6	0.7	0.8	0.9	1
p	1	0.97	0.89	0.78	0.66	0.55	0.44	0.34	0.25	0.15	0
R_{CTRL}	1	0.99	0.97	0.94	0.89	0.83	0.76	0.67	0.57	0.46	0.33
$R_{\text{SWAP-}x}$	0.83	0.83	0.82	0.79	0.76	0.73	0.68	0.63	0.58	0.53	0.5

For all values of ϵ, the attack causes no errors. However, in principle, it can be detected because it causes different loss rates to different types of bits: the loss rate experienced by Bob in the CTRL bits, R_{CTRL}, is usually different from the loss rate in the SWAP-x bits, $R_{\text{SWAP-x}}$ (see Table 6 for details). Therefore, in principle, the attack can be detected by a statistical test for most values of ϵ.

The loss rates become equal only for the value $\epsilon = \epsilon_0 \triangleq \sqrt{\frac{3-\sqrt{3}}{2}} \approx 0.796$ (which gives $\kappa^2 = \frac{\epsilon^2}{3}$). It seems that this specific attack *cannot* be detected, even in principle: it causes no errors, and it causes the same loss rate for all qubits. For this attack, Eve gets the information bit with probability $p = \frac{1}{4}$, and the loss rates are $R_{\text{CTRL}} = R_{\text{SWAP-x}} = \frac{1}{\sqrt{3}} \approx 0.577$. Therefore, this attack gives Eve a reasonable amount of information, and it is not detectable by looking at errors or comparing loss rates. (We can slightly modify the attack to make the loss rate the same in both directions of the quantum channel, too.)

We conclude that this weaker attack gives Eve partial information, causes no errors, and causes several loss rates. We also conclude that since the loss rates caused by the attack are usually different for different types of bits, the attack can be detected, in principle, for any value of ϵ except ϵ_0. However, for $\epsilon = \epsilon_0$, the attack seems undetectable.

5. Discussion

We have discussed a simpler and natural variant of the Mirror protocol (the "simplified Mirror protocol") which is easier to implement. We have found the simplified Mirror protocol to be completely non-robust; therefore, this protocol is actually an "over-simplified" Mirror protocol. We have presented in Section 4.1 an attack giving Eve full information without causing any error; in addition, since this attack also causes full loss of the CTRL bits, we have presented in Section 4.2 weaker attacks giving Eve partial information, causing no errors, and causing fewer losses. In particular, we have presented a specific attack (characterized by the parameter $\epsilon = \epsilon_0 \triangleq \sqrt{\frac{3-\sqrt{3}}{2}} \approx 0.796$) that seems undetectable and gives Eve one quarter ($\frac{1}{4}$) of all information bits.

Those attacks prove that the simplified Mirror protocol, which allows Alice to use only three classical operations (CTRL, SWAP-10, and SWAP-01), is completely non-robust. On the other hand, the Mirror protocol is proved completely robust (see Section 2 and [16]). As explained in Section 3, the only difference between the simplified Mirror protocol and the Mirror protocol is that the Mirror protocol allows a fourth classical operation, SWAP-ALL; therefore, allowing the SWAP-ALL operation is necessary for robustness. More generally, the Mirror protocol probably cannot be made much simpler while remaining robust: its complexity is crucial for robustness. Therefore, we have seen that if we want to use an SQKD protocol that is experimentally feasible in a secure way, we may have to use a relatively complicated protocol.

In this paper, we have not checked the experimental feasibility of Eve's attacks, because Eve is usually assumed to be all-powerful. Nonetheless, it can be interesting to check in the future the experimental feasibility of those attacks and discover whether the simplified Mirror protocol is flawed also in practice and not "only" in theory. Other interesting directions for future research include trying to find experimentally feasible SQKD protocols that are simpler than the Mirror protocol, and trying to find similar attacks against other QKD and SQKD protocols that have no complete robustness proof.

Author Contributions: T.M. suggested to investigate the robustness of the simplified protocol. M.B. suggested and designed the two attacks. All authors performed the careful analysis of the attacks, wrote the manuscript, and reviewed and commented on the final manuscript.

Funding: The work of Tal Mor and Rotem Liss was partly supported by the Israeli MOD Research and Technology Unit.

Acknowledgments: The authors thank Natan Tamari and Pavel Gurevich for useful discussions about the experimental implementation of SWAP-ALL.

Conflicts of Interest: The authors declare no conflict of interest.

Abbreviations

The following abbreviations are used in this manuscript:

QKD Quantum Key Distribution
SQKD Semiquantum Key Distribution

References

1. Boyer, M.; Kenigsberg, D.; Mor, T. Quantum Key Distribution with Classical Bob. *Phys. Rev. Lett.* **2007**, *99*, 140501. [CrossRef] [PubMed]
2. Boyer, M.; Gelles, R.; Kenigsberg, D.; Mor, T. Semiquantum key distribution. *Phys. Rev. A* **2009**, *79*, 032341. [CrossRef]
3. Zou, X.; Qiu, D.; Li, L.; Wu, L.; Li, L. Semiquantum-key distribution using less than four quantum states. *Phys. Rev. A* **2009**, *79*, 052312. [CrossRef]
4. Boyer, M.; Mor, T. Comment on "Semiquantum-key distribution using less than four quantum states". *Phys. Rev. A* **2011**, *83*, 046301. [CrossRef]
5. Lu, H.; Cai, Q.Y. Quantum key distribution with classical Alice. *Int. J. Quantum Inf.* **2008**, *06*, 1195–1202. [CrossRef]
6. Sun, Z.W.; Du, R.G.; Long, D.Y. Quantum key distribution with limited classical Bob. *Int. J. Quantum Inf.* **2013**, *11*, 1350005. [CrossRef]
7. Yu, K.F.; Yang, C.W.; Liao, C.H.; Hwang, T. Authenticated semi-quantum key distribution protocol using Bell states. *Quantum Inf. Process.* **2014**, *13*, 1457–1465. [CrossRef]
8. Krawec, W.O. Mediated semiquantum key distribution. *Phys. Rev. A* **2015**, *91*, 032323. [CrossRef]
9. Zou, X.; Qiu, D.; Zhang, S.; Mateus, P. Semiquantum key distribution without invoking the classical party's measurement capability. *Quantum Inf. Process.* **2015**, *14*, 2981–2996. [CrossRef]
10. Krawec, W.O. Security proof of a semi-quantum key distribution protocol. In Proceedings of the 2015 IEEE International Symposium on Information Theory (ISIT), Hong Kong, China, 14–19 June 2015, pp. 686–690. [CrossRef]
11. Krawec, W.O. Security of a semi-quantum protocol where reflections contribute to the secret key. *Quantum Inf. Process.* **2016**, *15*, 2067–2090. [CrossRef]
12. Zhang, W.; Qiu, D.; Mateus, P. Security of a single-state semi-quantum key distribution protocol. *Quantum Inf. Process.* **2018**, *17*, 135. [CrossRef]
13. Krawec, W.O. Practical security of semi-quantum key distribution. In *Proceedings of SPIE, Quantum Information Science, Sensing, and Computation X*; Donkor, E., Ed.; SPIE: Washington, DC, USA, 2018; Volumme 10660, p. 1066009. [CrossRef]
14. Tan, Y.G.; Lu, H.; Cai, Q.Y. Comment on "Quantum Key Distribution with Classical Bob". *Phys. Rev. Lett.* **2009**, *102*, 098901. [CrossRef] [PubMed]
15. Boyer, M.; Kenigsberg, D.; Mor, T. Boyer, Kenigsberg, and Mor Reply. *Phys. Rev. Lett.* **2009**, *102*, 098902. [CrossRef]
16. Boyer, M.; Katz, M.; Liss, R.; Mor, T. Experimentally feasible protocol for semiquantum key distribution. *Phys. Rev. A* **2017**, *96*, 062335. [CrossRef]
17. Brassard, G.; Lütkenhaus, N.; Mor, T.; Sanders, B.C. Limitations on Practical Quantum Cryptography. *Phys. Rev. Lett.* **2000**, *85*, 1330–1333. [CrossRef] [PubMed]

© 2018 by the authors. Licensee MDPI, Basel, Switzerland. This article is an open access article distributed under the terms and conditions of the Creative Commons Attribution (CC BY) license (http://creativecommons.org/licenses/by/4.0/).

Article

Probabilistic Teleportation of Arbitrary Two-Qubit Quantum State via Non-Symmetric Quantum Channel

Kan Wang [1,*], Xu-Tao Yu [2,*], Xiao-Fei Cai [2] and Zai-Chen Zhang [1]

1. National Mobile Communications Research Laboratory, Southeast University, Nanjing 210096, China; zczhang@seu.edu.cn
2. State Key Lab. of Millimeter Waves, Southeast University, Nanjing 210096, China; xiaofei.cai@nokia-sbell.com
* Correspondence: wangkan@seu.edu.cn (K.W.); yuxutao@seu.edu.cn (X.-T.Y.)

Received: 17 December 2017; Accepted: 28 March 2018; Published: 29 March 2018

Abstract: Quantum teleportation has significant meaning in quantum information. In particular, entangled states can also be used for perfectly teleporting the quantum state with some probability. This is more practical and efficient in practice. In this paper, we propose schemes to use non-symmetric quantum channel combinations for probabilistic teleportation of an arbitrary two-qubit quantum state from sender to receiver. The non-symmetric quantum channel is composed of a two-qubit partially entangled state and a three-qubit partially entangled state, where partially entangled Greenberger–Horne–Zeilinger (GHZ) state and W state are considered, respectively. All schemes are presented in detail and the unitary operations required are given in concise formulas. Methods are provided for reducing classical communication cost and combining operations to simplify the manipulation. Moreover, our schemes are flexible and applicable in different situations.

Keywords: quantum teleportation; entanglement; quantum channel; quantum communication

---

1. Introduction

Quantum teleportation, firstly proposed by Bennett et al. [1], is a feasible technique for moving quantum states via pre-established quantum channel amongst distant network nodes with the help of classical information. It is at the heart of many quantum information protocols and also represents a fundamental ingredient to the development of many quantum technologies, including quantum network [2,3], quantum secure communication [4,5], measurement-based quantum computing [6,7], and quantum repeater [8–10], etc. Due to its potential applications in the realm of quantum communication [11], a growing amount of theoretical and experimental progress [12–16] has been made in this domain.

As the transmission channel of teleportation, quantum entanglement [17] is fragile resource. The requirement of a maximally entangled quantum channel connecting nodes is very difficult to achieve or maintain in practice since the inevitable presence of noise reduces the entanglement of the quantum state shared between them. In practical implementations of the teleportation protocol, one can either adopt entanglement purification and distillation techniques to purify the states or use the partially entangled state to teleport quantum state perfectly with some probability. Probabilistic teleportation was introduced by Li et al. [18], following which several resources have used different types of entanglement. These are obtainable in [19–24].

Agrawal et al. [19] utilized a partially entangled state as a shared resource to teleport an unknown two-qubit state. Dai et al. [20,21] presented two protocols for probabilistically teleporting an arbitrary two-qubit state via two partially entangled W states and by the combination of a partially entangled

GHZ state and an entangled W state, respectively. Probabilistic teleportation of an arbitrary two-qubit entangled state can also be obtained via a dimensional four-qubit partially entangled cluster state by Xia et al. [22]. Liu et al. [23] proposed a teleportation protocol of an unknown two-qubit state probabilistically with partial information. Recently, Choudhury et al. [24] proposed protocol for probabilistic teleportation using POVM and projective measurement. Here, we focus on probabilistic teleportation schemes for an arbitrary two-qubit quantum state.

The existing works mostly use a symmetric quantum channel, i.e., two entangled states of two qubits or two entangled states of three qubits, to teleport two-qubit quantum state. However, the entangled states shared among quantum nodes in network would not be guaranteed to be of the same type. Different types of entanglements would be utilized as quantum channels. In this paper, we study the probabilistic teleportation using *non-symmetric quantum channel* for transmitting arbitrary two-qubit quantum state. The non-symmetric quantum channel consists of a two-qubit entangled state and a three-qubit entangled state. Schemes using different quantum channel combinations are proposed that could be seen as supplementary to the protocol family of teleporting two-qubit state. Methods are provided for reducing the classical communication cost and combining the separate unitary operations to simplify the whole process. Furthermore, in many existing protocols, the intermediate states and the unitary operations applied are shown in the form of complex tables. One of the unique features in this paper is that all unitary operations applied by the receiver and intermediate states in the process are summarized in concise formulas. With these formulas, the operations and intermediate states can be obtained through calculation rather than searching through complex tables.

The rest of this paper is organized as follows: Section 2 provides system model and quantum channels we considered in this paper. Section 3 discusses the probabilistic teleportation schemes using partially entangled GHZ state and two-qubit partially entangled state as quantum channel and the method for reducing classical communication cost. Another scheme is presented in Section 4 using another non-symmetric quantum channel combination (i.e., partially entangled W state and two-qubit partially entangled state). A method is given to combine unitary operations into one under the same basis as well. In Sections 5 and 6, we present a discussion and conclude the whole paper.

2. System Model

In this paper, we consider two nodes, conveniently called Alice and Bob, who share entangled states as quantum channel. Through the channel, Alice wants to transmit arbitrary two-qubit state to Bob as described below

$$|\chi\rangle = a_0|00\rangle + a_1|01\rangle + a_2|10\rangle + a_3|11\rangle, \quad (1)$$

where $a_i (i = 0, 1, 2, 3)$ is the amplitude of respective basis state satisfying the normalized condition $\sum_{i=0}^{3} |a_i|^2 = 1$. The quantum channel shared between two nodes consists of a two-qubit partially entangled state and three-qubit partially entangled state. GHZ state and W state are fundamental entangled states of three qubits, and widely used in protocols for transmitting quantum states. They represent diverse types of but can cover all three-qubit entangled states. Without losing generality, both GHZ state and W state are studied as part of quantum channel combination but separately in different schemes. The two-qubit partially entangled state and three-qubit partially entangled states are described as

$$|\psi\rangle = c|00\rangle + d|11\rangle, \quad \text{where } |c|^2 + |d|^2 = 1 \text{ and } |c| \geq |d|,$$
$$|GHZ\rangle = m|000\rangle + n|111\rangle, \quad \text{where } |m|^2 + |n|^2 = 1 \text{ and } |m| \geq |n|, \quad (2)$$
$$|W\rangle = x|001\rangle + y|010\rangle + z|100\rangle, \quad \text{where } |x|^2 + |y|^2 + |z|^2 = 1 \text{ and } |x| \geq |y| \geq |z|.$$

The system can be summarized into one general model as shown in Figure 1 where a two-qubit partially entangled state and a three-qubit partially entangled state are shared between Alice and Bob as a non-symmetric quantum channel. The classical communication channel is equipped.

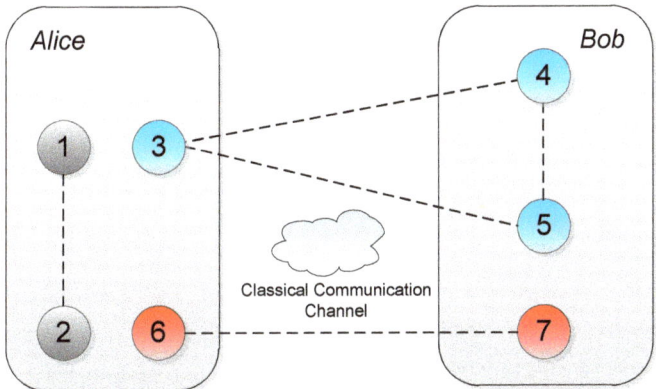

Figure 1. System model for teleporting arbitrary two-qubit state from Alice to Bob via non-symmetric quantum channel. For the convenience of description, we assume particles 1 and 2 are in the possession of Alice. Particle 3 from three-qubit partially entangled state is with Alice while Bob has particles 4 and 5. Particles 6 and 7 of two-qubit partially entangled state belong to Alice and Bob, respectively.

In the following paper, the three-qubit partially entangled W state is firstly considered as quantum channel together with two-qubit partially entangled state. We refer to this scheme as scheme A and the channel combination as non-symmetric quantum channel combination A. In addition, a special case for teleporting two-qubit entangled state is discussed using the same quantum channel combination as scheme A. Then, the three-qubit partially entangled GHZ state is used to replace former partially entangled W state. Similarly, we refer to it as scheme B and the non-symmetric quantum channel combination B. In the following, we present these schemes in detail and give methods for improvement.

3. Schemes Using Non-Symmetric Quantum Channel Combination A

In this section, the scheme to transmit an arbitrary two-qubit state is presented when using defined non-symmetric quantum channel combination A. As a special case of scheme A, teleportation of two-qubit entangled state is discussed accompanied with a method for reducing the classical communication cost in that case.

3.1. Teleporting Arbitrary Two-Qubit Quantum State

The non-symmetric quantum channel utilized for probabilistic teleportation of arbitrary two-qubit state is composed of a two-qubit partially entangled state and three-qubit partially entangled W state. Taking the states described in Equation (2), the initial system state can be written as

$$|\Phi_{sys}\rangle = |\chi\rangle_{12} \otimes |W\rangle_{345} \otimes |\psi\rangle_{67}. \tag{3}$$

To realize teleportation, the detailed process is elaborated as follows:

Step 1: Alice firstly performs two Bell-state measurements on particles (1, 3) and particles (2, 6), respectively. The system state may collapse into one of the 16 possible states, which can be expressed as $\langle\beta_{ij}|_{26}\langle\beta_{kl}|_{13} \Phi_{sys}\rangle$ in general, where $|\beta_{ij}\rangle$ and $|\beta_{kl}\rangle$ represent corresponding Bell states in the form

$$|\beta_{ij}\rangle = (|0,j\rangle + (-1)^i |1, 1-j\rangle)/\sqrt{2}, \tag{4}$$

where $i,j = 0,1$. Bell-state measurement results are expressed as classical bit strings $m_p m_q(|\beta_{ij}\rangle) \equiv ij$, where $m_p m_q$ denote the measurement results of particle p and q, respectively. Then, Alice sends these measurement results to Bob through classical communication channel immediately. According to the measurement results $m_1m_3m_2m_6$, the 16 possible states can be divided into four groups as follows, where $\overline{m_i}$ indicates the negation of the measurement outcome m_i.

A. When $m_1m_3m_2m_6$ is 0000, 0010, 1000 or 1010, i.e., $\overline{m_3}\,\overline{m_6}=1$, the system state is expressed as

$$\frac{1}{2}(a_0xc|010\rangle + a_0yc|100\rangle + (-1)^{m_1}a_2zc|000\rangle + (-1)^{m_2}a_1xd|011\rangle \\ + (-1)^{m_2}a_1yd|101\rangle + (-1)^{m_1\oplus m_2}a_3zd|001\rangle)_{457}. \tag{5}$$

B. When $m_1m_3m_2m_6$ is 0001, 0011, 1001 or 1011, i.e., $\overline{m_3}\,m_6=1$, the system state is expressed as

$$\frac{1}{2}(a_0xd|011\rangle + a_0yd|101\rangle + (-1)^{m_1}a_1xc|010\rangle + (-1)^{m_1}a_1yc|100\rangle \\ + (-1)^{m_2}a_2zd|001\rangle + (-1)^{m_1\oplus m_2}a_3zc|000\rangle)_{457}. \tag{6}$$

C. When $m_1m_3m_2m_6$ is 0100, 0110, 1100 or 1110, i.e., $m_3\,\overline{m_6}=1$, the system state is expressed as

$$\frac{1}{2}(a_0zc|000\rangle + (-1)^{m_2}a_1zd|001\rangle + (-1)^{m_1}a_2xc|010\rangle + (-1)^{m_1}a_2yc|100\rangle \\ + (-1)^{m_1\oplus m_2}a_3xd|011\rangle + (-1)^{m_1\oplus m_2}a_3yd|101\rangle)_{457}. \tag{7}$$

D. When $m_1m_3m_2m_6$ is 0101, 0111, 1101 or 1111, i.e., $m_3\,m_6=1$, the system state is expressed as

$$\frac{1}{2}(a_0zd|001\rangle + (-1)^{m_1}a_2xd|011\rangle + (-1)^{m_1}a_2yd|101\rangle + (-1)^{m_2}a_1zc|000\rangle \\ + (-1)^{m_1\oplus m_2}a_3xc|010\rangle + (-1)^{m_1\oplus m_2}a_3yc|100\rangle)_{457}. \tag{8}$$

Step 2: After receiving the classical information from Alice, Bob performs projective measurement on particle 4. If the result is $|0\rangle_4$ (denoted by $m_4=0$), the original state can not be reconstructed and the teleportation fails. Otherwise, Bob continues to apply following operations to recover the teleported quantum state.

Step 3: Then, for retrieving the correspondence between coefficients a_i and basis states, Bob needs to apply unitary operation U_{57} on particles (5, 7). The specific unitary operation required is determined by the measurement result according to the formula

$$U_{57} = (Z^{m_1}X^{\overline{m_3}})_5 \otimes (Z^{m_2}X^{m_6})_7, \tag{9}$$

where $X = \begin{bmatrix} 0 & 1 \\ 1 & 0 \end{bmatrix}$ and $Z = \begin{bmatrix} 1 & 0 \\ 0 & -1 \end{bmatrix}$ are Pauli matrices. After unitary operation U_{57}, the specific system state is determined by m_3m_6 and changes into

$$|\Phi'_{sys}\rangle = \begin{cases} a_0xc|00\rangle + a_1xd|01\rangle + a_2zc|10\rangle + a_3zd|11\rangle & \text{when } \overline{m_3}\,\overline{m_6}=1, \\ a_0xd|00\rangle + a_1xc|01\rangle + a_2zd|10\rangle + a_3zc|11\rangle & \text{when } \overline{m_3}\,m_6=1, \\ a_0zc|00\rangle + a_1zd|01\rangle + a_2xc|10\rangle + a_3xd|11\rangle & \text{when } m_3\,\overline{m_6}=1, \\ a_0zd|00\rangle + a_1zc|01\rangle + a_2xd|10\rangle + a_3xc|11\rangle & \text{when } m_3\,m_6=1. \end{cases} \tag{10}$$

Step 4: Bob introduces an auxiliary particle A with its initial state $|0\rangle_A$ and applies a collective unitary operation on particles (5, 7, A). To reconstruct the original state under the basis $\{|\beta_{ij}\rangle_{57}|0\rangle_A, |\beta_{ij}\rangle_{57}|1\rangle_A\}$ (where $|\beta_{ij}\rangle_{57}$ stands for the computational basis of an four-dimensional Hilbert space), the unitary operation should take the form

$$U_{57A} = \begin{pmatrix} C_1 & C_2 \\ C_2 & -C_1 \end{pmatrix}. \tag{11}$$

The $C_i(i=1,2)$ are 4×4 matrices in the form

$$\begin{cases} C_1 = diag(c_1, c_2, c_3, c_4), \\ C_2 = diag(\sqrt{1-c_1^2}, \sqrt{1-c_2^2}, \sqrt{1-c_3^2}, \sqrt{1-c_4^2}), \end{cases} \tag{12}$$

where c_i ($i=1,2,3,4$ and $|c_i| \leq 1$) and their corresponding C_i all depend on the specific system state. The specific form of c_i is summarized into the following expressions:

$$C_1 = \begin{cases} diag(\frac{zd}{xc}, \frac{z}{x}, \frac{d}{c}, 1) & \text{when } \overline{m_3}\,\overline{m_6} = 1, \\ diag(\frac{z}{x}, \frac{zd}{xc}, 1, \frac{d}{c}) & \text{when } \overline{m_3}\,m_6 = 1, \\ diag(\frac{d}{c}, 1, \frac{zd}{xc}, \frac{z}{x}) & \text{when } m_3\,\overline{m_6} = 1, \\ diag(1, \frac{d}{c}, \frac{z}{x}, \frac{zd}{xc}) & \text{when } m_3\,m_6 = 1. \end{cases} \quad (13)$$

Step 5: Finally, Bob performs projective measurement on particle A. The result $|1\rangle_A$ (denoted by $m_A = 1$) indicates the failure of this teleportation; on the contrary, if the result is $m_A = 0$, the two-qubit state has been reconstructed on particles 5 and 7, yielding a successful teleportation.

The success probability of scheme A is $4|zd|^2$. When $|x| = |y| = |z| = 1/\sqrt{3}$ and $|c| = |d| = 1/\sqrt{2}$, i.e., the quantum channel consists of two maximally entangled states, the success probability would reach its maximum $2/3$. The whole scheme is shown in Figure 2 and an example is given for illustrating the whole process better.

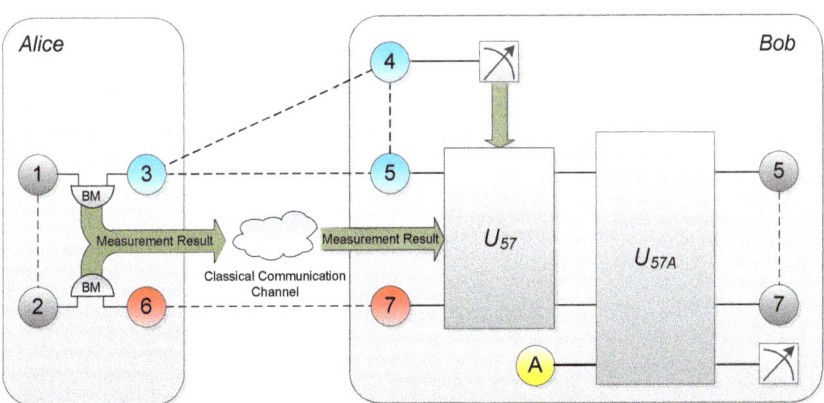

Figure 2. Probabilistic teleportation scheme utilizing partially entangled two-qubit state and W state as quantum channel. Bell-state measurements, projective measurements and local unitary operations are applied, together with the classical information, for original state recovery.

Example 1. *Assume the Bell-state measurement results $m_1 m_3 m_2 m_6 = 0000$. According to Equation (5), the system state after Alice's two Bell-state measurements should be*

$$\langle \beta_{00}|_{26} \langle \beta_{00}|_{13} \Phi_{sys} \rangle = \frac{1}{2}(a_0 xc|010\rangle + a_0 yc|100\rangle + a_2 zc|000\rangle + a_1 xd|011\rangle + yd|101\rangle + a_3 zd|001\rangle)_{457}.$$

Bob then measures particle 4. If the result is $m_4 = 1$, the teleportation fails. Otherwise, he continues to apply unitary operation $U_{57} = X_5 \otimes I_7$ on particles (5, 7) according to Equation (9). After that, Bob introduces an auxiliary particle A with its initial state $|0\rangle_A$ and the system state changes into

$$\left|\Phi''_{sys}\right\rangle = \frac{1}{2}(a_0 xc|000\rangle + a_1 xd|010\rangle + a_2 zc|100\rangle + a_3 zd|110\rangle)_{57A}.$$

Choosing $C_1 = diag(\frac{zd}{xc}, \frac{z}{x}, \frac{d}{c}, 1)$, corresponding eight-dimensional unitary operation U_{57A} is made up as below according to Equations (11) and (12).

$$U_{57A} = \begin{bmatrix} \frac{zd}{xc} & 0 & 0 & 0 & \sqrt{1-\left(\frac{zd}{xc}\right)^2} & 0 & 0 & 0 \\ 0 & \frac{z}{x} & 0 & 0 & 0 & \sqrt{1-\left(\frac{z}{x}\right)^2} & 0 & 0 \\ 0 & 0 & \frac{d}{c} & 0 & 0 & 0 & \sqrt{1-\left(\frac{d}{c}\right)^2} & 0 \\ 0 & 0 & 0 & 1 & 0 & 0 & 0 & 0 \\ \sqrt{1-\left(\frac{zd}{xc}\right)^2} & 0 & 0 & 0 & -\frac{zd}{xc} & 0 & 0 & 0 \\ 0 & \sqrt{1-\left(\frac{z}{x}\right)^2} & 0 & 0 & 0 & -\frac{z}{x} & 0 & 0 \\ 0 & 0 & \sqrt{1-\left(\frac{d}{c}\right)^2} & 0 & 0 & 0 & -\frac{d}{c} & 0 \\ 0 & 0 & 0 & 0 & 0 & 0 & 0 & -1 \end{bmatrix}.$$

After the operation, Bob obtains the system state

$$|\Phi_{final}\rangle = U_{57A}|\Phi''_{sys}\rangle = \frac{zd}{2}(a_0|00\rangle_{57} + a_1|01\rangle_{57} + a_2|10\rangle_{57} + a_3|11\rangle_{57})|0\rangle_A$$
$$+ \left(\sqrt{x^2c^2-z^2d^2}a_0|00\rangle_{57} + d\sqrt{x^2-z^2}a_1|01\rangle_{57} + z\sqrt{c^2-d^2}a_2|10\rangle_{57}\right)|1\rangle_A.$$

Then, undertaking measurements on particle A by Bob, the result $m_A=0$ indicates that the original state has been recovered at Bob (on particles 5 and 7) successfully while $m_A=1$ means failure of teleportation. Using a two-qubit partially entangled state and a three-qubit partially entangled W state as quantum channel, we could teleport arbitrary two-qubit state probabilistically.

3.2. Teleporting Two-Qubit Entangled State

In this subsection, we apply scheme A on teleportation of two-qubit entangled state. In addition, we introduce a new method of processing classical information that will help to reduce the classical communication cost needed in this case. Without losing generality, the two-qubit entangled state to be teleported is assumed in the general form $|\gamma\rangle = \alpha|00\rangle + \beta|11\rangle$, where $|\alpha|^2 + |\beta|^2 = 1$ and $|\alpha| \geq |\beta| > 0$. This form is widely used in related works and any entangled two-qubit state can be brought to this form via local unitary operations. Alice still performs Bell-state measurements on particles (1, 3) and particles (2, 6), respectively. The system state would collapse into one of the 16 possible states, which are classified according to measurement results as follows:

$$|\Phi'_{sys}\rangle = \begin{cases} \frac{1}{2}(\alpha xc|010\rangle + \alpha yc|100\rangle + (-1)^{m_1 \oplus m_2}\beta zd|001\rangle)_{457} & \text{when } \overline{m_3}\,\overline{m_6} = 1, \\ \frac{1}{2}(\alpha xd|011\rangle + \alpha yd|101\rangle + (-1)^{m_1 \oplus m_2}\beta zc|000\rangle)_{457} & \text{when } \overline{m_3}\,m_6 = 1, \\ \frac{1}{2}(\alpha zc|000\rangle + (-1)^{m_1 \oplus m_2}\beta xd|011\rangle + (-1)^{m_1 \oplus m_2}\beta yd|101\rangle)_{457} & \text{when } m_3\,\overline{m_6} = 1, \\ \frac{1}{2}(\alpha zd|001\rangle + (-1)^{m_1 \oplus m_2}\beta xc|010\rangle + (-1)^{m_1 \oplus m_2}\beta yc|100\rangle)_{457} & \text{when } m_3\,m_6 = 1. \end{cases} \quad (14)$$

The projective measurement on particle 4 is also performed by Bob after receiving the classical information from Alice. When measurement result of particle 4 is $m_4 = 0$, Bob continues to apply unitary operation U_{57} on particles (5, 7) as

$$U_{57} = (X^{\overline{m_3}})_5 \otimes (Z^{m_1 \oplus m_2} X^{m_6})_7. \quad (15)$$

An auxiliary particle A is introduced with its initial state $|0\rangle_A$, and then the collective unitary operation U_{57A} is constructed by Bob. The U_{57A} takes the same form with Equations (11)–(13) because the unitary operation is only related with the quantum channel characters. Measurement on particle A is still required so that Bob can judge whether this teleportation succeeds or not. The total success probability of teleportation is $4|zd|^2$.

Comparing these two schemes, the main difference is the formulation of U_{57}, which is closely related with the Bell-state measurement results sent from Alice. The original four cbits information $m_1 m_3 m_2 m_6$ is sent to Bob directly through classical communication channel. However, after observing Equation (15), we get the conclusion that whether the Z operation on particle 7 is required or not is determined by the XOR result between m_1 and m_2. Thus, we use $m_x = m_1 \oplus m_2$ to denote the XOR result and the Equation (15) can be rewritten as

$$U_{57} = (X^{\overline{m_3}})_5 \otimes (Z^{m_x} X^{m_6})_7. \tag{16}$$

Through this combination, only the XOR result m_x instead of the respective measurement results of particles 1 and 2 needs to be sent to Bob together with $m_3 m_6$. These 3 cbits information rather than 4 cbits are enough for Bob to determine U_{57} so that the classical communication cost is reduced by 25%, which is one of the advantages of this scheme.

Remark 1. *Actually, if setting the parameter $a_1 = a_2 = 0$ in Equation (1), we can also get the case discussed above. Scheme A for teleporting arbitrary two-qubit state is a more general one where the scheme for teleporting two-qubit entangled state can be seen as a special case. The method presented for reducing the classical communication cost is only valid in this case. If the two-qubit entangled state is given in other forms, the expression of U_{57} would change accordingly. However, for avoiding re-derivation, Alice could use the result-mapping method proposed in [25] to obtain the operation U_{57} correctly.*

4. Scheme Using Non-Symmetric Quantum Channel Combination B

In the schemes above, the maximal success probability of teleportation could only reach 2/3 even if the quantum channel is composed of corresponding maximally entangled states. Analyzing the schemes, the measurement on particle 4 is the main source of the non-fully recoverable system state. In this section, the three-qubit partially entangled GHZ state is utilized in quantum channel combination B and the corresponding scheme using such quantum channel to complete arbitrary two-qubit state teleportation is presented. The initial system state is expressed as

$$|\Phi_{sys}\rangle = |\chi\rangle_{12} \otimes |GHZ\rangle_{345} \otimes |\psi\rangle_{67}, \tag{17}$$

where $|GHZ\rangle_{345}$ is partially entangled GHZ state with the form as Equation (2). Similar to aforementioned schemes, Alice performs two Bell-state measurements on particles (1, 3) and particles (2, 6), respectively. The system state collapses and can be divided into four groups as follows:

A. When $m_1 m_3 m_2 m_6$ is 0000, 0010, 1000 or 1010, i.e., $\overline{m_3}\,\overline{m_6} = 1$, the system state is expressed as

$$\frac{1}{2}\left(a_0 mc|000\rangle + (-1)^{m_2} a_1 md|001\rangle + (-1)^{m_1} a_2 nc|110\rangle + (-1)^{m_1 \oplus m_2} a_3 nd|111\rangle\right)_{457}. \tag{18}$$

B. When $m_1 m_3 m_2 m_6$ is 0001, 0011, 1001 or 1011, i.e., $\overline{m_3} m_6 = 1$, the system state is expressed as

$$\frac{1}{2}\left(a_0 md|001\rangle + (-1)^{m_2} a_1 mc|000\rangle + (-1)^{m_1} a_2 nd|111\rangle + (-1)^{m_1 \oplus m_2} a_3 nc|110\rangle\right)_{457}. \tag{19}$$

C. When $m_1 m_3 m_2 m_6$ is 0100, 0110, 1100 or 1110, i.e., $m_3 \overline{m_6} = 1$, the system state is expressed as

$$\frac{1}{2}\left(a_0 nc|110\rangle + (-1)^{m_2} a_1 nd|111\rangle + (-1)^{m_1} a_2 mc|000\rangle + (-1)^{m_1 \oplus m_2} a_3 md|001\rangle\right)_{457}. \tag{20}$$

D. When $m_1 m_3 m_2 m_6$ is 0101, 0111, 1101 or 1111, i.e., $m_3 m_6 = 1$, the system state is expressed as

$$\frac{1}{2}\left(a_0 nd|111\rangle + (-1)^{m_2} a_1 nc|110\rangle + (-1)^{m_1} a_2 md|001\rangle + (-1)^{m_1 \oplus m_2} a_3 mc|000\rangle\right)_{457}. \tag{21}$$

The main difference of this scheme is that Bob needs to send particle 4 through additional Hadamard gate ($H = \frac{1}{\sqrt{2}} \begin{bmatrix} 1 & 1 \\ 1 & -1 \end{bmatrix}$) before measurement as shown in Figure 3. Afterwards, Bob performs projective measurement on particle 4. The unitary operation U_{57} performed on particles (5, 7) to retrieve the correspondence would be determined by the measurement result m_4 together with $m_1 m_3 m_2 m_6$ through the following formula:

$$U_{57} = (-1)^{m_3 m_4} (Z^{m_1 \oplus m_4} X^{m_3})_5 \otimes (Z^{m_2} X^{m_6})_7. \tag{22}$$

The system state changes into

$$|\Phi'_{sys}\rangle = \begin{cases} \frac{1}{2\sqrt{2}} (a_0 mc |00\rangle + a_1 md |01\rangle + a_2 nc |10\rangle + a_3 nd |11\rangle)_{57} & \text{when } \overline{m_3}\,\overline{m_6} = 1, \\ \frac{1}{2\sqrt{2}} (a_0 md |00\rangle + a_1 mc |01\rangle + a_2 nd |10\rangle + a_3 nc |11\rangle)_{57} & \text{when } \overline{m_3}\, m_6 = 1, \\ \frac{1}{2\sqrt{2}} (a_0 nc |00\rangle + a_1 nd |01\rangle + a_2 mc |10\rangle + a_3 md |11\rangle)_{57} & \text{when } m_3 \overline{m_6} = 1, \\ \frac{1}{2\sqrt{2}} (a_0 nd |00\rangle + a_1 nc |01\rangle + a_2 md |10\rangle + a_3 mc |11\rangle)_{57} & \text{when } m_3 m_6 = 1. \end{cases} \tag{23}$$

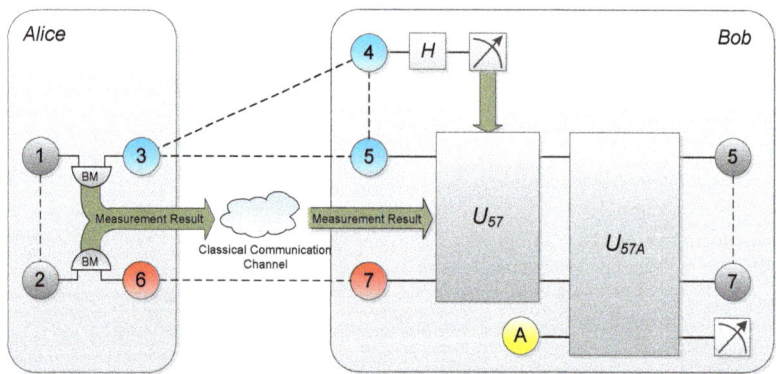

Figure 3. Probabilistic teleportation scheme utilizing partially entangled two-qubit state and GHZ state as quantum channel. Extra H operation is applied before operation U_{57} to avoid failure of teleportation.

Then, Bob introduces an auxiliary particle A and performs a collective unitary operation U_{57A} on particles (5, 7, A) to correct the distortion on system state and recover the original state. The transformation matrix of U_{57A} should be constructed in accordance with Equations (11) and (12). The exact form of C_1 is also determined by $m_3 m_6$ and expressed as

$$C_1 = \begin{cases} diag(\frac{nd}{mc}, \frac{n}{m}, \frac{d}{c}, 1) & \text{when } \overline{m_3}\,\overline{m_6} = 1, \\ diag(\frac{n}{m}, \frac{nd}{mc}, 1, \frac{d}{c}) & \text{when } \overline{m_3}\, m_6 = 1, \\ diag(\frac{d}{c}, 1, \frac{nd}{mc}, \frac{n}{m}) & \text{when } m_3 \overline{m_6} = 1, \\ diag(1, \frac{d}{c}, \frac{n}{m}, \frac{nd}{mc}) & \text{when } m_3 m_6 = 1. \end{cases} \tag{24}$$

After operation U_{57A}, projective measurement is performed on particle A. Similarly, only the result $m_A = 0$ indicates successful teleportation. The total success probability is $4|nd|^2$. If $|m| = |n| = |c| = |d| = 1/\sqrt{2}$, i.e., the quantum channel is composed of two maximally entangled states, the success probability would reach its maximum of 1. With the help of partially entangled GHZ state and additional H operation, scheme B could increase the maximal success probability by avoiding failure occurring after measurement on particle 4.

In the above schemes, Bob applies two separate unitary operations under different bases. U_{57} is constructed under computational basis while U_{57A} is made under the basis of $\{|\beta_{ij}\rangle_{57}|0\rangle_A, |\beta_{ij}\rangle_{57}|1\rangle_A\}$. These two unitary operations can not be combined directly through tensor product or matrix multiplication. For combining them into one operation under unified basis, we introduce the transformation matrix T between these two bases. With this matrix, U_{57} and U_{57A} could be combined into one complete unitary operation under computational basis. The detailed method is shown in the following part and the basis transformation matrix T is given as

$$T = \begin{bmatrix} 1 & 0 & 0 & 0 & 0 & 0 & 0 & 0 \\ 0 & 0 & 1 & 0 & 0 & 0 & 0 & 0 \\ 0 & 0 & 0 & 0 & 1 & 0 & 0 & 0 \\ 0 & 0 & 0 & 0 & 0 & 0 & 1 & 0 \\ 0 & 1 & 0 & 0 & 0 & 0 & 0 & 0 \\ 0 & 0 & 0 & 1 & 0 & 0 & 0 & 0 \\ 0 & 0 & 0 & 0 & 0 & 1 & 0 & 0 \\ 0 & 0 & 0 & 0 & 0 & 0 & 0 & 1 \end{bmatrix}. \tag{25}$$

Based on the existing unitary operation U_{57A} in Equation (22), the new unitary operation $\hat{U}_{57A}$ under computational basis can be obtained through $\hat{U}_{57A} = T^{-1} U_{57A} T$, where the matrix T is equivalent to transition matrix actually. Hence, Bob could introduce the auxiliary particle $|0\rangle_A$ after obtaining the measurement result of particle 4 but before performing unitary operation. With the newly constructed unitary operation matrix $\hat{U}_{57A}$, the two unitary operations can be combined into one operation U' under computational basis easily in the form of

$$U' = \hat{U}_{57A}(U_{57} \otimes I_A), \tag{26}$$

where I_A is a two-dimensional identity matrix. We illustrate this method by taking $m_1 m_3 m_2 m_6 m_4 = 01110$ as an example. According to Equation (21), after measurement on particle 4 and introducing auxiliary particle $|0\rangle_A$, the system state should be

$$\left|\Phi''_{sys}\right\rangle = \frac{1}{2\sqrt{2}}(-a_3 mc|000\rangle + a_2 md|010\rangle - a_1 nc|100\rangle + a_0 nd|110\rangle)_{57A}, \tag{27}$$

which can be expressed with vector under computational basis as

$$\left|\Phi''_{sys}\right\rangle = \frac{1}{2\sqrt{2}}[-a_3 mc, 0, a_2 md, 0, -a_1 nc, 0, a_0 nd, 0]^T.$$

Then, perform the combined unitary operation U'

$$U' = \hat{U}_{57A}(U_{57} \otimes I_A) = T^{-1}\begin{pmatrix} C_1 & C_2 \\ C_2 & -C_1 \end{pmatrix} T(X_5 \otimes (ZX)_7 \otimes I_A), \tag{28}$$

where $C_1 = \text{diag}(1, \frac{d}{c}, \frac{n}{m}, \frac{nd}{mc})$ accordingly so that the final system state is

$$\left|\Phi_{final}\right\rangle = U'\left|\Phi''_{sys}\right\rangle = \frac{1}{2\sqrt{2}}[a_0 nd, 0, a_1 nd, a_1 n\sqrt{c^2 - d^2}, a_2 nd, a_2 d\sqrt{m^2 - n^2}, a_3 nd, a_3 \sqrt{c^2 m^2 - d^2 n^2}]^T$$

$$= \frac{dn}{2\sqrt{2}}(a_0|00\rangle + a_1|01\rangle + a_2|10\rangle + a_3|11\rangle)_{57}|0\rangle_A \tag{29}$$

$$+ \frac{1}{2\sqrt{2}}\left(a_1 n\sqrt{c^2 - d^2}|01\rangle + a_2 d\sqrt{m^2 - n^2}|10\rangle + a_3\sqrt{c^2 m^2 - d^2 n^2}|11\rangle\right)_{57}|1\rangle_A.$$

Similarly, we need measurement on auxiliary particle A. The measurement result $m_A = 0$ indicates successful teleportation. On the contrary, $m_A = 1$ suggests failure.

Remark 2. *Obviously, only one unitary operation is performed instead of the previous two separate operations, which would simplify the quantum manipulation. Less operation may lead to reduction in the possibility of making an error in practice. This method can also be applied in our previous presented schemes compatibly, and the whole teleportation process may benefit from simplified operation. In addition, when the quantum channel is composed of maximally entangled states, the success probability may reach its maximum. In that situation, the system state after operation U_{57} had already been recovered to its original state successfully. There is no need to introduce auxiliary particle and apply operation U_{57A} any more.*

5. Discussion

We present schemes utilizing two different non-symmetric quantum channel combinations to teleport arbitrary two-qubit state probabilistically in this paper. One scheme consists of two-qubit partially entangled state and three-qubit partially entangled W state, and the other one consists of two-qubit partially entangled state and three-qubit partially entangled GHZ state. We still refer to these two schemes as scheme A and scheme B in the following discussion and conclusion.

(1) The belonging of particle 4

In the paper, we assume that particle 4 from the three-qubit partially entangled state is held by Bob. However, the belonging of the particle should be flexible. There are still two different situations in addition to what we have considered. We will analyze them case by case to show that our schemes are also applicable in the situations where Bob does not have particle 4. (a) When Alice has particle 4, Bob only has particles 5 and 7. In this situation, Alice should perform the measurement on particle 4 after the Bell-state measurements in scheme A or after H operation in scheme B, and then send Bob the result. If Alice gets the result $m_4 = 1$ in scheme A, she should stop the whole teleportation process and restart another one immediately. Otherwise, if the result is $m_4 = 0$, Alice should send measurement results to Bob, and then Bob continues to apply the corresponding operation for recovering the original state. (b) When particle 4 belongs to a third party Charles, the system changes to a controlled teleportation model where the measurement on particle 4 should be performed by Charles. As a trusted third party, Charles can control the whole teleportation because Bob cannot recover the original state and complete the teleportation without Charles' cooperation (measurement and its result). In a real application, the reasonable allocation of particle 4 should be determined according to specific condition and purpose. Our presented schemes can work well in all three of these three situations with few modifications.

(2) Another understanding of our schemes

In both schemes, Alice needs to perform two Bell-state measurements on particles firstly and send measurement results to Bob through classical communication channels. Then, Bob performs the measurement on particle 4. If a partially entangled GHZ state is utilized, an extra H operation should be applied before Bob's measurement. After that, an auxiliary particle is introduced for reconstructing original state by applying specific unitary operation(s) according to their measurement results.

If analyzed from another point of view, the measurement on particle 4 of three-qubit partially entangled W state in scheme A and operations (H operation and measurement) on particle 4 of three-qubit partially entangled GHZ state can be seen as the process of preparing a two-qubit entangled state for the teleportation protocol. The remainder of the teleportation can then be analyzed using previously explored techniques [26]. We will refer to this as Scheme R for the discussion below.

This intuitive understanding of the protocol may help us to explain why maximum success probability of scheme A using W state can not reach 1. This is primarily because a two-qubit partially entangled state can only be obtained with some probability from the remainder of the two-qubit system of the W state after measurement. The probability is exactly 2/3, which limits the maximum of success probability of the whole teleportation. The result of measurement on particle 4 indicates whether the two-qubit entangled channel is prepared successfully. In contrast, we can obtain a two-qubit partially entangled state with certainty from the operations on particle 4 of the partially entangled GHZ state so that the maximum success probability of scheme B can reach 1.

Comparing scheme R with our schemes, the main difference is the order of performing measurement on particle 4 while other manipulations are similar [26]. In scheme R, when Alice intends to teleport the two-qubit state, she needs to notify Bob to perform measurement on particle 4 to initiate the whole teleportation process. Then, Bob sends back information to Alice with notification that the channel is ready so that Alice can perform the Bell measurement and follow-up steps using the prepared two two-qubit entangled states as the channel. Obviously, there are two additional classical communication processes compared with our schemes. Extra classical communication cost for protocol control and more transmission delay will be introduced especially in the system model discussed in this paper.

However, if Alice holds particle 4 of the three-qubit partially entangled state, the additional communication processes of scheme R are unnecessary. Alice does not need to ask Bob to perform the measurement: instead she could do the preparation herself and be aware of whether the channel is ready. In addition, Alice could terminate the follow-up steps and restart another teleportation if she failed to get the two-qubit partially entangled state from the W state (when the result is 1), which would increase the whole success probability.

Based on the above analysis, our schemes proposed in paper may avoid introducing extra cost for the whole teleportation process. They are still meaningful and can be applied in specific scenes when necessary. This provides one feasible choice of teleportation scheme for two-qubit state transmitting.

6. Conclusions

In future quantum networks, the quantum states to be teleported and the entangled states shared among nodes are diverse. We studied probabilistic teleportation of two-qubit quantum states using partially entangled states as a channel in this paper. Two schemes were presented using non-symmetric quantum channels, which is different from the existing work. The quantum channel consists of a two-qubit partially entangled state and a three-qubit partially entangled state. Both GHZ state and W state were considered as representatives of three-qubit entangled states to give more complete solutions. The composite quantum channel we discussed may exist in real applications. To some extent, our schemes are supplementary to the protocol family of two-qubit state teleportation.

In addition, we illustrated the whole teleportation process in detail and the unitary operations required were given in concise formulas rather than tables. The required operation can be worked out through calculation instead of searching through complex tables, which is helpful in fast automatic control and processing. A method for reducing the classical communication cost in the special case of teleporting two-qubit entangled state was provided. By sending only 3 cbits compressed Bell-state measurement outcome to the receiver instead of 4 cbits, as in other related works, the cost could be reduced by 25%. Finally, the transformation matrix T was provided for converting unitary operation under different bases used. With matrix T, the former two separate unitary operations under different bases can be combined into one under unified computational basis. Less operation is associated with simpler manipulation and reduced possibility of error in theory. Furthermore, our schemes are applicable in other situations where there is some flexibility regarding where the particle belongs, as we discussed.

Schemes using partially entangled states to realize probabilistic teleportation are crucial for the practical application of quantum communication and networks. We hope our work may stimulate more investigations into proposals for quantum communication and networks. In future work, we plan to study the teleportation in the presence of unavoidable noises and test the efficiency of the protocol.

Acknowledgments: This project was supported by the National Science Foundation of China (No.61601120); the China Postdoctoral Science Foundation (No.2016M591742); and the Jiangsu Planned Projects for Postdoctoral Research Funds (No.1601166C).

Author Contributions: K.W. and X.T.Y. conceived and designed these schemes; K.W. and X.F.C. performed the derivation and calculation; K.W. wrote the paper; Z.C.Z. reviewed the draft and gave helpful suggestions.

Conflicts of Interest: The authors declare no conflict of interest.

References

1. Bennett, C.H.; Brassard, G.; Crépeau, C.; Jozsa, R.; Peres, A.; Wootters, W.K. Teleporting an unknown quantum state via dual classical and Einstein-Podolsky-Rosen channels. *Phys. Rev. Lett.* **1993**, *70*, 1895–1899.
2. Yonezawa, H.; Aoki, T.; Furusawa, A. Demonstration of a quantum teleportation network for continuous variables. *Nature* **2004**, *431*, 430–433.
3. Wang, K.; Yu, X.T.; Lu, S.L.; Gong, Y.X. Quantum wireless multihop communication based on arbitrary Bell pairs and teleportation. *Phys. Rev. A* **2014**, *89*, 022329.
4. Zhang, W.; Ding, D.S.; Sheng, Y.B.; Zhou, L.; Shi, B.S.; Guo, G.C. Quantum Secure Direct Communication with Quantum Memory. *Phys. Rev. Lett.* **2017**, *118*, 220501.
5. Deng, F.G.; Long, G.L.; Liu, X.S. Two-step quantum direct communication protocol using the Einstein-Podolsky-Rosen pair block. *Phys. Rev. A* **2003**, *68*, 042317.
6. Raussendorf, R.; Briegel, H.J. A One-Way Quantum Computer. *Phys. Rev. Lett.* **2001**, *86*, 5188–5191.
7. Brassard, G.; Braunstein, S.L.; Cleve, R. Teleportation as a quantum computation. *Phys. D Nonlinear Phenom.* **1998**, *120*, 43–47.
8. Briegel, H.J.; Dür, W.; Cirac, J.I.; Zoller, P. Quantum Repeaters: The Role of Imperfect Local Operations in Quantum Communication. *Phys. Rev. Lett.* **1998**, *81*, 5932–5935.
9. Sangouard, N.; Simon, C.; de Riedmatten, H.; Gisin, N. Quantum repeaters based on atomic ensembles and linear optics. *Rev. Mod. Phys.* **2011**, *83*, 33–80.
10. Van Meter, R.; Ladd, T.D.; Munro, W.J.; Nemoto, K. System Design for a Long-line Quantum Repeater. *IEEE/ACM Trans. Netw.* **2009**, *17*, 1002–1013.
11. Gisin, N.; Thew, R. Quantum communication. *Nat. Photonics* **2007**, *1*, 165.
12. Bouwmeester, D.; Pan, J.W.; Mattle, K.; Eibl, M.; Weinfurter, H.; Zeilinger, A. Experimental quantum teleportation. *Nature* **1997**, *390*, 575–579.
13. Braunstein, S.L.; Kimble, H.J. Teleportation of Continuous Quantum Variables. *Phys. Rev. Lett.* **1998**, *80*, 869–872.
14. Ma, X.S.; Herbst, T.; Scheidl, T.; Wang, D.; Kropatschek, S.; Naylor, W.; Wittmann, B.; Mech, A.; Kofler, J.; Anisimova, E.; et al. Quantum teleportation over 143 kilometres using active feed-forward. *Nature* **2012**, *489*, 269–273.
15. Pirandola, S.; Eisert, J.; Weedbrook, C.; Furusawa, A.; Braunstein, S.L. Advances in quantum teleportation. *Nat. Photonics* **2015**, *9*, 641.
16. Wang, X.L.; Cai, X.D.; Su, Z.E.; Chen, M.C.; Wu, D.; Li, L.; Liu, N.L.; Lu, C.Y.; Pan, J.W. Quantum teleportation of multiple degrees of freedom of a single photon. *Nature* **2015**, *518*, 516–519.
17. Horodecki, R.; Horodecki, P.; Horodecki, M.; Horodecki, K. Quantum entanglement. *Rev. Mod. Phys.* **2009**, *81*, 865–942.
18. Li, W.L.; Li, C.F.; Guo, G.C. Probabilistic teleportation and entanglement matching. *Phys. Rev. A* **2000**, *61*, 034301.
19. Agrawal, P.; Pati, A.K. Probabilistic quantum teleportation. *Phys. Lett. A* **2002**, *305*, 12–17.
20. Dai, H.Y.; Chen, P.X.; Li, C.Z. Probabilistic teleportation of an arbitrary two-particle state by two partial three-particle entangled W states. *J. Opt. B Quantum Semiclassical Opt.* **2004**, *20*, 106.
21. Dai, H.Y.; Chen, P.X.; Li, C.Z. Probabilistic teleportation of an arbitrary two-particle state by a partially entangled three-particle GHZ state and W state. *Opt. Commun.* **2004**, *231*, 281–287.
22. Xia, Y.; Song, J.; Song, H.S. Classical communication help and probabilistic teleportation with one-dimensional non-maximally entangled cluster states. *Int. J. Theor. Phys.* **2008**, *47*, 1552–1558.
23. Liu, D.; Huang, Z.; Guo, X. Probabilistic Teleportation via Quantum Channel with Partial Information. *Entropy* **2015**, *17*, 3621–3630.

24. Choudhury, B.S.; Dhara, A. Probabilistically teleporting arbitrary two-qubit states. *Quantum Inf. Process.* **2016**, *15*, 5063–5071.
25. Wang, K.; Gong, Y.X.; Yu, X.T.; Lu, S.L. Addendum to "Quantum wireless multihop communication based on arbitrary Bell pairs and teleportation". *Phys. Rev. A* **2014**, *90*, 044302.
26. Yong-Jian, G.; Yi-Zhuang, Z.; Guang-Can, G. Probabilistic Teleportation of an Arbitrary Two-particle State. *Chin. Phys. Lett.* **2006**, *18*, 1543.

© 2018 by the authors. Licensee MDPI, Basel, Switzerland. This article is an open access article distributed under the terms and conditions of the Creative Commons Attribution (CC BY) license (http://creativecommons.org/licenses/by/4.0/).

MDPI
St. Alban-Anlage 66
4052 Basel
Switzerland
Tel. +41 61 683 77 34
Fax +41 61 302 89 18
www.mdpi.com

Entropy Editorial Office
E-mail: entropy@mdpi.com
www.mdpi.com/journal/entropy

www.ingramcontent.com/pod-product-compliance
Lightning Source LLC
LaVergne TN
LVHW070546100526
838202LV00012B/394